COLLOID CHEMISTRY

Original title

KOLOIDNA KHIMIYA

The first edition was published in Bulgarian by
the State Publishing House for Science and Art, Sofia, Bulgaria,
in two volumes, 1957 and 1958.

A revised edition was published in Russian by
the State Publishing House for Foreign Literature, Moscow, 1960.

This new revised English edition was translated by
EXPRESS TRANSLATION SERVICE, LONDON, S.W. 19.

COLLOID CHEMISTRY

by

A. SHELUDKO, D.Sc.

Professor of Physical Chemistry,
University of Sofia, Bulgaria

ELSEVIER PUBLISHING COMPANY

AMSTERDAM – LONDON – NEW YORK

1966

ELSEVIER PUBLISHING COMPANY
335 JAN VAN GALENSTRAAT, P.O. BOX 211, AMSTERDAM

AMERICAN ELSEVIER PUBLISHING COMPANY, INC.
52 VANDERBILT AVENUE, NEW YORK, N.Y. 10017

ELSEVIER PUBLISHING COMPANY LIMITED
RIPPLESIDE COMMERCIAL ESTATE, BARKING, ESSEX

LIBRARY OF CONGRESS CATALOG CARD NUMBER 65–12561

WITH 70 ILLUSTRATIONS AND 12 TABLES

PRINTED IN THE NETHERLANDS

Preface to the third edition

The present book is based on a somewhat amplified course of lectures on colloid chemistry with which I have been entrusted since 1955 at the Faculty of Chemistry of Sofia University.

The present state of colloid chemistry is characterised by an abundance of experimental material, whilst the corresponding theoretical background is relatively limited and uncoordinated. For this reason in the syllabus of most universities the main stress is laid on the facts, and any theoretical explanations, especially those of a quantitative character, are only given sporadically wherever this is feasible, and as far as this is justified by the abundance of experimental material.

My university experience has shown, however, that it is the theory which presents the greatest difficulty, and that, given a theoretical grounding, the student is in a better position to absorb and understand the experimental data, which he can also independently assimilate in the course of practical work. On the other hand, the recent development of colloid chemistry already makes it possible to formulate, if not a comprehensive, then at least a coherent, theory of the physical chemistry of microheterogeneous systems, which underlies the study of colloids.

In planning the present course I therefore only preserved the bare minimum of facts necessary to illustrate the theory, and completely discarded certain very important sections which still lack a more or less comprehensive theoretical treatment, such as the study of the rheological properties of colloids.

In this I have followed the example of the excellent but unfortunately rather obsolete analogous work by Zhukov [*Colloid Chemistry* (Kolloidnaya khimiya), Leningrad University, 1949], and of the broader exposition given by Overbeek and Jonker in Kruyt's *Colloid Science* (Elsevier Publishing Company, 1952).

It is hoped that the large number of available textbooks and papers, some of which are quoted in this book and which are devoted largely to the experimental aspects of this branch of chemistry, will enable the interested reader to supplement his knowledge in this respect.

In the present edition the tendency towards generalization has been carried a little further. Thus for instance the thermodynamics of heterogeneous systems, kinetics of the formation of new phases, and some other problems of the physical chemistry of surfaces have been expanded into a separate chapter. Even this procedure does not do justice to the role of surface phenomena and physical chemistry of microheterogeneous systems, so that, as in the case of electrokinetic phenomena, independent sections are also devoted to thermodynamics of heterogeneous systems, adsorption, and phenomena of the formation of new phases.

A special feature of this edition is the greater attention paid to thin layers, as yet mostly from the thermodynamic point of view. I believe that this field is at present the key to the further development of colloid science. Only the future can show whether this is in fact the case, and it will then be possible to put forward a colloid course in a more final and complete form.

There is no doubt in my mind that attempts at generalization should be made, even when they are rather controversial and incomplete, since this will contribute to the evolution of colloid chemistry from its qualitative descriptive state to a more rigorous discipline based on quantitative analysis of the observed phenomena.

Even after this second revision, the book lacks completeness, as regard both the choice of the optimum proportions of experimental material and theory, and the coordination of individual parts.

When working on the first Bulgarian edition of this book I received considerable assistance from Academician R. Kaishev. The second (Russian) edition was enlarged with substantial collaboration of the Corresponding Member of the Academy of Sciences of the USSR, B. V. Deryagin, and of Professor S. S. Voyutskii.

I wish to take this opportunity of expressing my gratitude to I. Ivanov, S. Stoilov, B. Mutaftchiev, Professors L. M. Shcherbakov, Yu. M. Glazman and S. S. Dukhin, Dr. H. Sonntag, and to many other colleagues who assisted me in the preparation of this edition.

The final English text was thoroughly discussed with Professor J. Th. G. Overbeek and especially with Professor J. Lyklema, who made numerous valuable suggestions. I wish to express my deepest gratitude to them for their help, and would add that for technical reasons it proved impossible to incorporate all of their suggestions in the printed text. Any omissions or errors are therefore entirely the responsibility of the author.

Sofia, Bulgaria, February 1966 A. SHELUDKO

Transliteration

Although in general a standard system has been used when transliterating the names of Russian authors, the reader will find that some names appear in two, or even three, different forms. This occurs when mention is made of a Russian author's work published in, for example, Germany or France. The transliteration system of the country in question is then used to facilitate reference. Every effort has been made to avoid errors but the reader's indulgence is asked if mistakes have occurred due to difficulties in checking sources.

Contents

CHAPTER I

Preparation and Purification of Lyophobic Colloidal Systems

Colloidal systems may be defined as systems containing particles smaller than 1 μ. They are called *lyophobic* when the system cannot be made by dissolution of the colloidal material in the solvent. The solvent is called the *dispersion medium*, and it is most frequently an aqueous solution. The dispersed particles are collectively known as the *disperse phase*. The finer the particles the greater the *degree of dispersion*. Each particle consists of a large number of molecules and is, consequently, a separate phase with a phase volume and phase boundary.

Systems in which the particle size exceeds 1 μ are called *suspensions*. Suspensions, especially the coarse ones, settle rapidly, whereas colloidal solutions are stable in this respect. Particles whose size is equal to 1 μ (this corresponds to the upper limit of the degree of dispersion of colloids) settle very slowly; when in water, such spherical particles whose specific gravity is greater than 1 settle only slowly, the sedimentation rate being of the order of 10^{-4} cm/sec.

The properties of the colloidal solutions of the compounds of high-molecular weight approach those of true molecular solutions. Examples of such systems are the solutions of proteins, rubber, and various synthetic polymerisation and polycondensation products. In these cases, although fully molecular dispersion is attained on dissolution, the molecules themselves are so large as to impart to these solutions a number of properties typical of colloids. Such solutions are sometimes called the lyophilic colloids, since they are more stable than the lyophobic colloids; this points to a considerably greater affinity between these substances and the solvent.

The present chapter contains a brief analysis of various procedures used for the preparation of lyophobic colloids.

In 1909, Svedberg[1] suggested a division of these methods into condensation and dispersion methods. In the former case, starting from molecular solutions, conditions are created in which individual molecules combine to form aggregates. In the case of dispersion methods, on the other hand, one phase is dispersed in another.

In a number of cases, *e.g.* when colloidal solutions are prepared by the dispersion of electrodes in an electric arc, both condensation and dispersion processes take place simultaneously, and it has not up to now been possible to establish which of these two processes does in fact predominate.

1. Condensation Methods

The most important factors to be observed in condensation methods are the production of nuclei, their growth, and finally their stabilisation. It is known that nuclei of a new phase form only in supersaturated solutions. Examples of supersaturated systems can be supersaturated solutions, supercooled vapours etc. Thus *e.g.* in the Wilson chamber the cloud forms during the adiabatic expansion and the corresponding supercooling of vapour-saturated air. The supersaturation indispensable for the formation of a nucleus in a solution, is obtained by the formation—by a chemical process—of products sparingly soluble in the given solvent, or by adding to the solution of the given substance another component which reduces its solubility, *e.g.* by adding water to the solution of sulphur in ethanol.

In principle, colloidal solutions can be obtained from all reactions associated with precipitation. To achieve this end, it is necessary to carry out the reaction under conditions at which the growth of particles during precipitation possesses a certain limit. The augmentation of the nuclei is due to both their growth in the supersaturated system and to the agglomeration of individual particles, the latter process being known as coagulation. Thus, a slow rate of formation of new nuclei and rapid aggregation give rise to a small number of large particles; in contrast, when the rate of formation of the nuclei is high and the growth of aggregates is slow, a colloidal solution containing a large number of small particles is obtained. It is obvious that the conditions of the process should prevent the possibility of a simultaneous intensive coagulation.

According to Volmer[2], the rate of formation of the nuclei may be expressed by

$$J = k_1 \exp\left[-\frac{k_2}{(\ln c/c_0)^2}\right], \tag{1.1}$$

References p. 11

where c is the concentration of the substance in the solution, and c_0 the equilibrium concentration*.

The rate of particle growth is usually determined by diffusion of substance towards these particles, and is proportional to the difference $(c - c_0)$. In some cases the deposition of the substance on particles is impeded by other factors, when the rate of particle growth may decrease substantially.

The formation of highly-dispersed precipitates is assisted by increasing the concentration of the reactants, in other words, by supersaturating the solution, since in this case the rate of formation of the nuclei increases faster than the growth rate of the particles formed. However, the increased supersaturation results at the same time in an increase of the total number of particles; it will be demonstrated in the more detailed analysis of the process of coagulation that this increase brings about an increase in the rate of particle aggregation. Therefore, for each reaction associated with the formation of a new disperse phase, there exist certain determined optimum conditions for obtaining the colloidal system. These optimum conditions are usually chosen in an empirical manner, since it is difficult to determine them theoretically.

Every factor which inhibits coagulation assists in the formation of a colloidal system. One of the very important conditions is a very low electrolyte content, since electrolytes increase the rate of coagulation even if present in low concentrations. This is why precipitation reactions, in which no electrolytes are formed, are particularly suitable for the preparation of colloidal solutions. The following reactions may be used as examples:

$$2H_2S + SO_2 \rightarrow 3S + 2H_2O,$$

$$As_2O_3 + 3H_2S \rightarrow As_2S_3 + 3H_2O.$$

An example of an unfavourable reaction can be:

$$AgNO_3 + KBr \rightarrow AgBr + KNO_3.$$

Some substances have the property of reducing the rate of coagulation. Silver or bromide ions exert such an effect on silver bromide. Consequent-

* For details concerning the question of the formation of a new phase see p. 101 *ff*, where instead of the expression $\ln(c/c_0)$, which measures supersaturation, the equivalent expression $\ln(p_r/p_\infty)$ is used.

ly it is advantageous to employ excess $AgNO_3$ or excess KBr in order to prepare stable AgBr. Such substances are called the *stabilising electrolytes.*

Practical indications relating to the preparation of colloidal solutions can be obtained if one analyses the techniques employed in analytical chemistry (gravimetric analysis) to prevent the formation of precipitate suspensions. However, the aim of the colloid chemist should be, in contrast, not the separation of the precipitate, but its preservation in a highly-dispersed state as a suspension in the solution.

When using chemical methods, one usually obtains *polydisperse* colloidal systems whose particles have various sizes. *Monodisperse* sols sometimes play a very important part in the study of the properties of colloidal systems; their preparation is based on the control of processes of the formation of nuclei and of their growth. When nuclei form, there appear particles of various size, mainly very small. If the subsequent growth of these particles occurs under identical conditions, the individual mass increments are identical. If these increments are large compared with the mass of the nuclei, the difference between the masses of the grown crystal particles or tiny droplets will be only small. Such is, *e.g.* the method of obtaining a monodisperse gold sol, suggested by Zsigmondy (1906)[3] and based on the reduction of gold with hydrogen peroxide or formaldehyde:

$$2HAuCl_4 + 3H_2O_2 \rightarrow 2Au + 8HCl + 3O_2,$$

$$2HAuCl_4 + 3HCHO + 11KOH \rightarrow 2Au + 3HCOOK + 8KCl + 8H_2O$$

This process is carried out in two stages: initially a so-called nuclear sol with submicroscopic particles is obtained and subsequently a weak supersaturation is produced in this sol, owing to which no new nuclei are formed, whereas the existing nuclei grow in size.

Colloidal solutions can also be obtained by the condensation of vapours. Such a method was suggested by Roginskii and Shal'nikov (1943)[4]; the apparatus used for this purpose is shown in Fig. 1.

Another approach has been used by la Mer and coworkers (1948) to obtain monodisperse sulphur sols. Addition of acid to a dilute solution of thiosulphate leads to the slow production of sulphur, which at the very outset is dissolved. As the reaction proceeds, the solution suddenly becomes sufficiently supersaturated to form nuclei. Once these are formed, no further new nucleation occurs because all newly formed sulphur will precipitate on the growing particles; consequently a monodisperse sol is formed (see e.g. Zaiser and La Mer[5]).

References p. 11

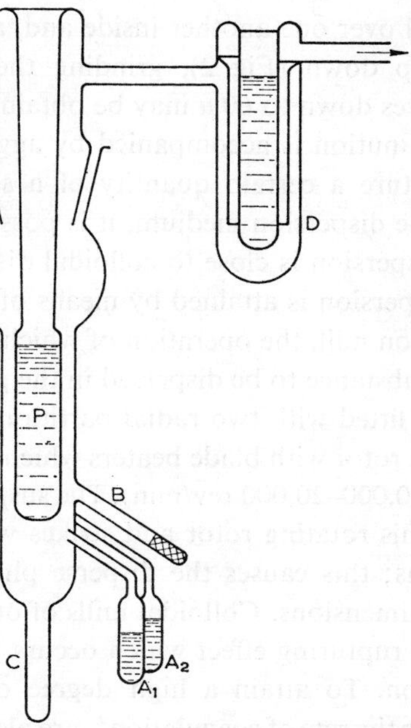

Fig. 1. Roginskii and Shal'nikov's apparatus for the production of colloidal solutions.

In this apparatus, containers A_1 and A_2 are filled with the solvent, and B contains the substance to be dispersed. A_1 and A_2 are immersed in liquid air, and air is evacuated from the vessel through the liquid-air trap D. Liquid air is then poured into the inner vessel P, and the side-pieces are heated. The ratio between the solvent and the substance to be dispersed (both of which condense on the vessel P) is determined by the temperature ratio of the side-pieces. Liquid air is then removed from P and the solid mixture which has formed on its walls liquefies and drops as a colloidal solution into C. This method enables very pure colloidal solutions of a great variety of substances to be obtained, *e.g.* the purple sols of sodium, blue-green sols of potassium and rubidium, brown-green sols of cesium in ether, benzene and other organic liquids.

2. Dispersion Methods

According to these methods, the substance is comminuted by using mills of various types, ball mills being used most frequently. When these mills rotate, the balls roll over one another inside and, at a certain position of the mill, they drop down (Fig. 2), grinding the preliminarily wetted mixture. Particle sizes down to 60 μ may be obtained. This limit is due to the fact that comminution is accompanied by agglomeration. If we add to the ground mixture a certain quantity of a soluble inert substance which will act as the dispersion medium, it is possible to obtain particles whose degree of dispersion is close to colloidal dispersion (1 μ).

A much finer dispersion is attained by means of the so-called colloidal mills, *e.g.* the Plauson mill, the operation of which is as follows. A coarse suspension of the substance to be dispersed in the given solvent is fed into a cylindrical stator fitted with two radial partitions which do not extend to the bottom and a rotor with blade beaters which rotates in this stator at a very high speed (10,000–20,000 rev/min). The suspension is carried along by the beaters of this rotating rotor and strikes with great force against the stator partitions; this causes the disperse phase to be comminuted down to colloidal dimensions. Colloidal mills of other types are based on the principle of the rupturing effect which occurs when a fast-moving jet changes its direction. To attain a high degree of dispersion peptizing agents, which reduce the rate of coagulation*, are also used with these mills.

3. Dispersion by Ultrasound

Ultrasonic vibrations are mechanical vibrations whose frequency exceeds that of audible sound (15,000–20,000 c/s). When ultrasound is used as a dispersing agent, the frequencies applied are roughly the same as those used in radio engineering, *i.e.* hundreds of thousands and millions of cycles per sec. These high-frequency vibrations were discovered and investigated by Langevin. Apart from a highly interesting physiological effect (*e.g.* killing of organisms) these ultrasonic waves can bring about an intensive colloidal comminution of one phase inside another. Thus, if a coarse suspension is subjected to sufficiently strong ultrasonic vibrations, it undergoes a dispersion down to colloidal dimensions.

* Rebinder and co-workers found that the surface-active substances used as additives during comminution not only increase the stability of the dispersion obtained, but also facilitate the dispersion itself by penetrating into microfissures which form in the deformed material and thus contribute to the destruction of the crystalline solid phase. This phenomenon is known as the "Rebinder effect" and is widely used in engineering.

References p. 11

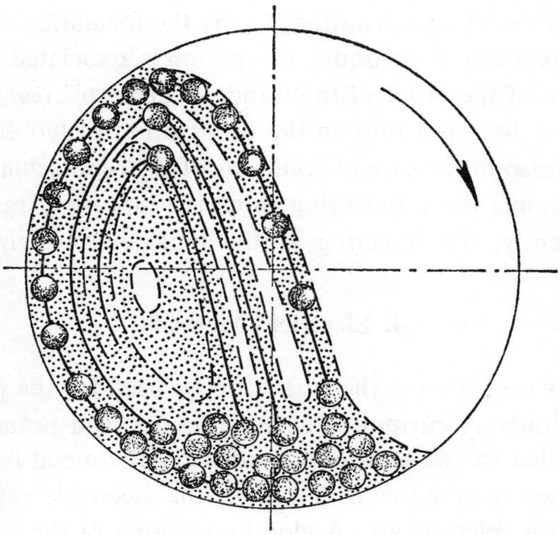

Fig. 2. A ball mill.

Ultrasonic vibrations may be obtained by two methods.

(a) A suitably cut quartz plate contracts and expands when acted upon by a variable electric field (this phenomenon is the reverse of the piezoelectric effect). If a high-frequency alternating current is supplied to such a plate the latter begins to oscillate synchronously with the electric vibrations. When the frequency of electric vibrations coincides with the self-frequency of vibrations of the quartz plate, the vibrations of the latter become particularly intensive. The quartz plate is placed in a non-conducting liquid (*e.g.* transformer oil) which transmits the vibrations; the liquid layer below the plate swells up and appears to boil. The test tube with the liquid mixture to be dispersed is placed in this vibrating liquid.

(b) Besides the quartz generator, ultrasound can also be generated by the so-called magnetostriction generators. Their operation is based on the ability of ferromagnetic materials to change their form in a magnetic field. A bar (or tube) of specified dimensions, made of nickel or a nickel–cobalt alloy and placed in the mixture to be dispersed, is exposed to a variable magnetic field of a certain frequency, which generates resonance vibrations in the bar. These generators are mostly used for generating vibrations of lower frequency.

The mechanism of the effect of ultrasound on colloidal systems has not yet been fully elucidated. A particularly important role in this case is played most probably by cavitation, *i.e.* by the formation of bubbles due to the local expansion of the liquid. Cavitation is associated with very high local pressures, of the order of thousands of atmospheres.

Ultrasound is used not only in the preparation of lyophobic colloidal solutions, but also in the case of solutions of high-molecular compounds, when the ultrasonic vibrations bring about the rupture of macromolecules and, consequently, the lowering of the molecular weight of the high polymer.

4. Electrical Methods

These methods were among the earliest to be used for the preparation of lyophobic colloids. A current of a few amperes at a potential of about 100 V is supplied to two electrodes made from the metal to be dispersed, immersed in water or aqueous solution, the electrode gap being varied so as to obtain an electric arc. A cloud consisting of the colloid-disperse metal or its oxide forms near the electrodes. This method was suggested by Bredig (1898)[6], but it suffers from the disadvantage that the dispersion is associated with an intense heating of the solution; therefore it cannot be used to obtain dispersions in organic liquids which decompose at high temperatures. Moreover, during the preparation of aqueous colloidal solutions the process is accompanied by intensive electrolysis, which results in the formation of secondary products.

These disadvantages have been removed to a certain extent in the method of Svedberg who suggested feeding the arc with a high-frequency alternating current from an induction coil.

5. Purification of Lyophobic Colloidal Systems

Colloidal solutions often contain soluble low-molecular components which must be eliminated for some reason. Thus, *e.g.*, electrolytes as a rule reduce the stability of sols and must be removed after a colloidal solution has been prepared. The general principle of purification of colloidal systems from the molecularly dissolved substances is based on the considerable difference in size between colloidal particles and molecules and on the ability of the latter to pass through fine-porous membranes which are impermeable to the colloidal phase. In general, lyophobic colloidal systems may be purified by the same methods as solutions of high-molecular substances.

References p. 11

Dialysis

As early as 1861 Graham[7] suggested the use of semi-permeable membranes for detecting the colloidal nature of a solution and for the purification of colloidal solutions by means of dialysis. In the process of dialysis the colloidal solution is usually placed in a vessel, the bottom or walls of which consist of semi-permeable membranes, and which is immersed in the pure solvent. The outer vessel is either very large or the solvent is exchanged continuously. The low-molecular components, which pass across the membrane, are fairly accurately eliminated from the solution.

The choice of membranes depends on the system concerned. Collodion or cellophane membranes are used most frequently in the case of aqueous solutions. Particularly convenient are collodion membranes, which can be easily prepared in the laboratory; by changing the conditions of preparation it is possible to obtain membranes with pores of various sizes. Animal membranes and parchment may also be used. The latter was used in Graham's investigations.

The rate of dialysis is very low, but increases with increasing area of the membrane; for this reason the membranes of various dialysers are folded. The rate of dialysis may also be increased by stirring the solutions.

Ultrafiltration

This method of purification consists of dialysis under pressure, the

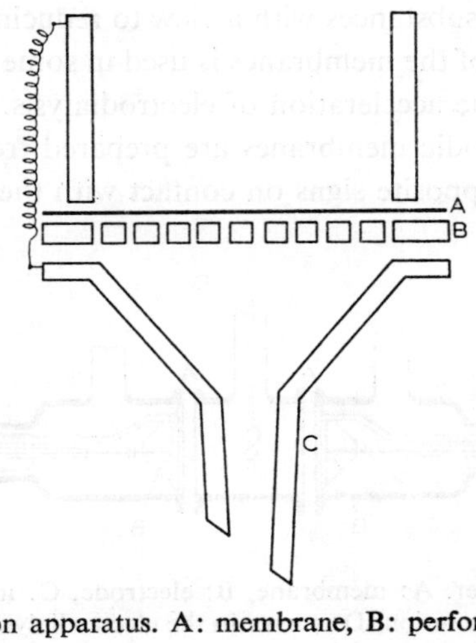

Fig. 3. Ultrafiltration apparatus. A: membrane, B: perforated plate, C: funnel.

dialysed solution being forced through the membrane. Ultrafiltration enables the colloidal solutions to be concentrated until complete separation of the disperse phase is achieved. To help the membrane to withstand the pressure, a reinforcing perforated plate of suitable material is placed underneath (Fig. 3).

Electrodialysis

When the low-molecular component is an electrolyte, the dialysis can be accelerated by applying an electric field to the auxiliary electrodes in the solution on both sides of the membrane. This method is particularly effective when the electrolyte to be separated is present in low concentrations. In these cases, the difference between the concentration of the electrolyte on both sides of the membrane is rather small, and therefore in the absence of an electric field the dialysis is very slow. Besides, the low conductance of these solutions enables strong electric fields to be used.

The diagram of an electrodialyser is shown in Fig. 4. Since the membranes acquire an electric charge on contact with the liquid, electrodialysis can be associated with changes in the ionic composition of the colloidal solution, and its pH can undergo a corresponding change. These changes are caused by the fact that the electrically charged membranes have an unequal permeability with respect to cations and anions. To avoid this undesirable effect the membranes to be used in the electrodialysis are treated with various substances with a view to reducing their own charge. The selective power of the membranes is used in some cases for a selective purification or for the acceleration of electrodialysis. In such a case, the anodic and the cathodic membranes are prepared from materials which acquire charges of opposite signs on contact with the solution.

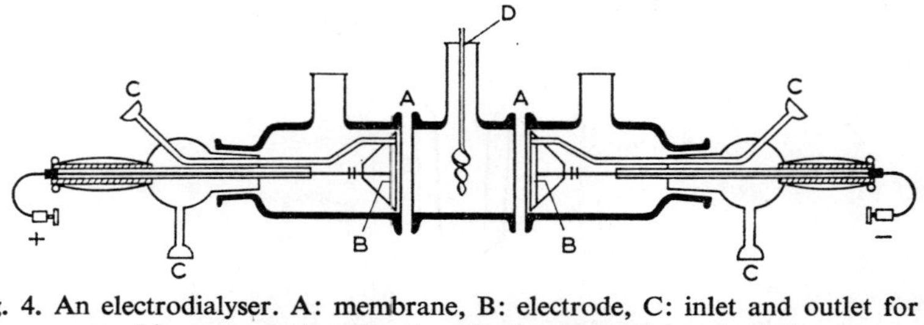

Fig. 4. An electrodialyser. A: membrane, B: electrode, C: inlet and outlet for the washing-out solution, D: stirrer in the electrodialysed solution.

References p. 11

REFERENCES

1 Svedberg, *Herstellung kolloïder Lösungen anorganischer Stoffe*, Dresden, 1909.
2 Volmer, *Die chemische Reaktion. Vol. IV. Kinetik der Phasenbildung*, Steinkopf, Leipzig, 1939.
3 Zsigmondy, *Z. Phys. Chem. (Leipzig)*, 56 (1906) 65.
4 Haritonov and Shal'nikov, *Mechanisms of condensation and formation of colloids*, G.T.T.I., Moscow, 1943.
5 Zaiser and La Mer, *J. Colloid Sci.*, 3 (1948) 571.
6 Bredig, *Z. Elektrochem.*, 4 (1898) 514, 547.
7 Graham, *Phil. Trans. Roy. Soc. London*, 151 (1861) 183.

CHAPTER II

The Optical Properties of Colloids

As far as their physical nature is concerned, the optical properties of colloidal systems present one of the most clearly understood sections of colloid science. However, at the same time, in view of the complexity of mathematical interpretation, the theory of these phenomena presents serious difficulties even to specialists in mathematical physics. It is due to such difficulties that many problems relating to this theory have not as yet been solved. Practical utilisation of a number of phenomena which have been explained in principle is very difficult and frequently even impossible, without resort to approximations which are not always reliable, or to fast electronic computers which are still not easily available.

In spite of these difficulties the science of the optical properties of colloids is in the course of rapid development owing to the high practical importance of this field, particularly in high-molecular systems. We therefore face the almost impossible task of giving a sufficiently complete and clear account of these highly complex questions and consequently, any success we might have, despite all our efforts must necessarily be limited. We shall try to explain qualitatively the fundamentals of the phenomena involved. As regards the quantitative aspect of the question, this must of necessity be confined mainly to the available formulae and their application to the simplest cases. In his further studies the reader is referred to monographs which have recently been published as a result of the ever increasing importance of this field of science. G. H. Jonker's exposition of the problem, adduced in Ref. 1 is brief, easily understandable and more complete than that presented here.

1. The Scattering of Light

One of the most important and clearly expressed optical properties of colloids consists in their pronounced ability to scatter light. This opales-

References p. 35

cence is sometimes called the Tyndall effect, after the investigator who studied it in detail (1868)[2].

During its propagation in any material medium, the electromagnetic wave of light polarises atoms and molecules composing this medium *i.e.* their centres of negative and positive charge are shifted with respect to each other. This polarisation follows the frequency of the illuminating light beam, thus causing the polarised structural elements of the substance to become themselves sources of luminescence. In a completely homogeneous medium, such as an ideal crystal, the diffuse light thus obtained produces interference according to the principle of Huyghens so that the lateral light is completely extinguished, and the beam propagates without any weakening (scattering). However, when the polarisability of the micro-regions of the medium is not the same everywhere, then the scattered light does not undergo a complete extinction, and opalescence appears. With the gradual increase of the dimensions of the inhomogeneous regions or their concentration it is possible to achieve a transition to the homogeneous system and, at the same time, the scattering of light should disappear.

The polarisability of a substance is governed by its dielectric constant ε or, according to the known relation,

$$\varepsilon = n^2 \tag{2.1}$$

by its refractive index n. Consequently, any inhomogeneity with respect to ε (or n) is a source of opalescence. Such inhomogeneity occurs, in the first place, in heterogeneous disperse systems of a substance whose refractive index differs from that of the dispersion medium. In molecular systems (liquids, gases) there also arise continuously, owing to fluctuations of density in one-component systems and fluctuations of concentration in multicomponent systems, micro-regions where n differs from its mean value, and therefore such media do scatter light.

(*a*) *Scattering by particles*

In the following discussion we shall first make the simplifying assumption that no consumptive light absorption occurs. This is in fact an important practical case since it includes all colourless systems.

When developing the theory of opalescence in the simplest way we study the scattering of light by one particle and generalise the obtained result for the totality of particles. The simplest case is that of scattering by a spherical particle, since it is only for complete symmetry of this kind that the scattering does not depend on the orientation of the particle with respect to the plane formed by the incident beam and by the direction of observation. If the particles do not possess the same diameter in different directions, it is necessary to take into account the scattering at all positions with respect to the plane referred to above. Given a random distribution all positions are equally probable and, after the mean value has been evaluated, the scattering is described again by the formula for a spherical particle* with a certain effective radius. If, however, for any reason the anisodimensional particles undergo an orientation, then the mean scattering by a particle in the case of complete orientation is described by a formula which corresponds to its form and orientation and, in the case of partial orientation, by a formula which corresponds to a certain effective form.

Generalisation of the result for the totality of particles can be achieved in the simplest way if their concentration is sufficiently low: one then obtains the result by simple summation for the whole number of particles. At higher concentrations the problem becomes more complicated, because the light diffused by individual particles undergoes interference and is partly extinguished; the particles are illuminated not only directly by the external source of illumination, but also by the diffuse light from other particles (multiple scattering); the light which falls on the particle is weakened by the scattering on other particles. As a result of these complex interrelations the resultant scattering is weaker than might be expected from simple additivity. Modern theory usually avoids these complex situations and only analyses the case of extremely dilute systems. It may be easily seen that the criterion of sufficient dilution is the proportionality between opalescence and concentration.

Thus, if we confine ourselves to dilute solutions, the problem is reduced to the study of scattering by one particle. Moreover, for the time being we can confine ourselves to spherical particles, particularly as in the absence of orientation this will also enable the scattering by sufficiently small anisodimensional particles to be described, if we assume for them a certain effective radius.

A further important simplification consists in assuming that the linear dimensions of the particle are small compared with the wavelength of the

* If the particles are sufficiently small.

References p. 35

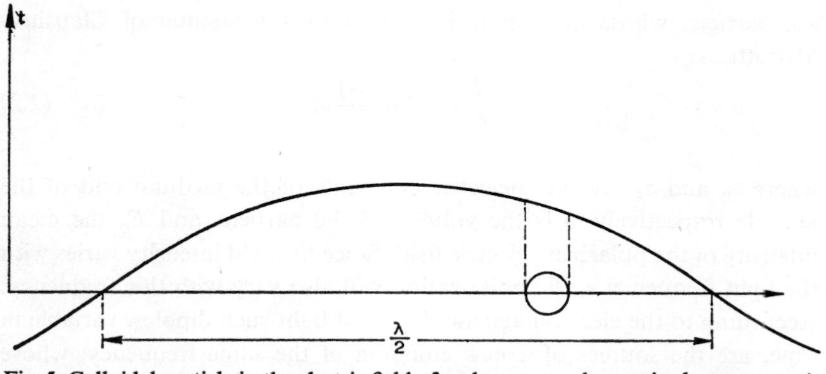

Fig. 5. Colloidal particle in the electric field of a plane monochromatic electromagnetic wave.

illuminating light. In this case the particle is confined in a space where the electromagnetic field force is at every moment almost the same (Fig. 5) along the entire length of the particle. It is frequently permissible to make use of this simplification, since in many cases the dimensions of particles in conventional colloidal systems are indeed small compared with the wavelength of visible light (of the order of 0.5 μ). This also applies to the high-molecular systems, and to the fluctuations of density and concentration.

The simplifications referred to above make it possible to attain the formulation of the problem, utilized by Rayleigh[3], who had based on it the first quantitative theory of light-scattering by a turbid medium*.

All molecules which form an isotropic spherical particle are polarised simultaneously and equally (homogeneous field) by the electromagnetic field of the illuminating beam. This results in the total polarisation P of

* Rayleigh's principal task was to explain the blue colour of the sky. To this end he first developed a theory of light scattering by particles (1871)[3], finding that brightness of the diffuse light varied inversely as the fourth power of the wavelength of light. Consequently, when white light is scattered, the scattered light is enriched in components of shorter wavelength thus acquiring a bluish tinge which is also characteristic of many colloidal solutions in the case of lateral illumination. On the other hand, transmitted light is enriched in components of longer wavelength and acquires a reddish tinge. Later (1899)[3] Rayleigh assumed that it is not particles but density fluctuations that are the main source of light scattering in the atmosphere; he developed a corresponding molecular theory of scattering by fluctuations, which was borne out by many measurements and enabled the Avogadro number to be determined with an accuracy of up to 1%. In this case the dependence on the light wavelength is the same.

the particle, whose magnitude is derived by the method of Clausius–Mosotti as:

$$P = \frac{3}{4\pi} \frac{\varepsilon_0(\varepsilon_1 - \varepsilon_0)}{\varepsilon_1 + 2\varepsilon_0} vE_0, \tag{2.2}$$

where ε_0 and ε_1 are the dielectric constants of the medium and of the particle respectively, v is the volume of the particle, and E_0 the mean intensity of the polarizing electric field. Since the field intensity varies with the light frequency ν, the polarization will also vary with this frequency. According to the electromagnetic theory of light such dipoles, variable in time, are the sources of a new emission of the same frequency, whose intensity I is given by the formula:

$$I = \frac{4\pi^3 \nu^4}{c_0^3 r^2} \sqrt{\varepsilon_0} P^2 \sin^2\varphi, \tag{2.3}$$

where φ is the angle between the direction of radiation and the dipole axis, c_0 the velocity of light in vacuum, and $r(r \gg \lambda)$ the distance from the observer. The field intensity E_0 is determined by the intensity I_0 of the illuminating beam as follows [the Umov–Poynting vector, which also underlies the derivation of eq. (2.3)]:

$$I_0 = \frac{c_0 \sqrt{\varepsilon_0}}{4\pi} E_0^2. \tag{2.4}$$

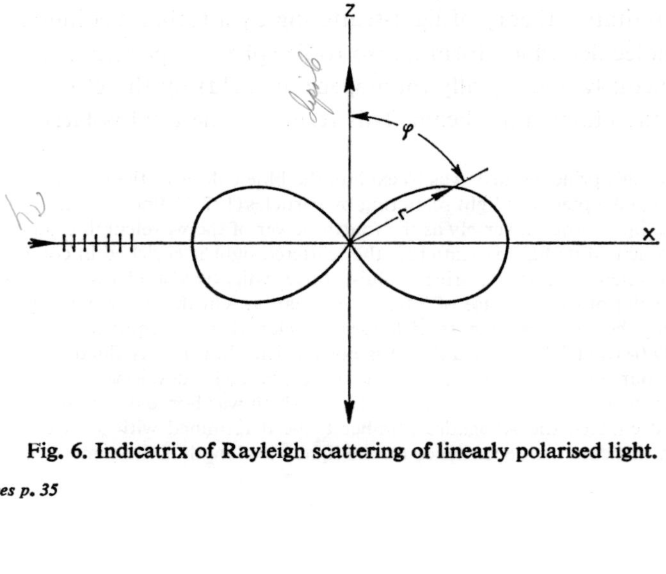

Fig. 6. Indicatrix of Rayleigh scattering of linearly polarised light.

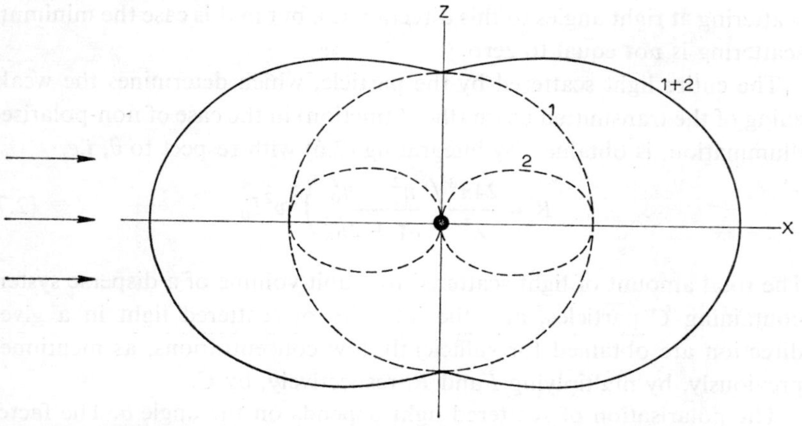

Fig. 7. Polar diagram of the intensity of Rayleigh scattering in the case of irradiation with a non-polarised beam. (1) Intensity of polarised radiation whose electric vector is at right angles to the plane of drawing. (2) Intensity of polarised radiation whose electric vector lies in the plane of the drawing. (1 + 2) total intensity.

By combining (2.1), (2.2), (2.3) and (2.4) we obtain the Rayleigh formula

$$I = \frac{9\pi^2 \sin^2\varphi}{r^2\lambda^4}\left(\frac{n_1^2 - n_0^2}{n_1^2 + 2n_0^2}\right)^2 v^2 I_0, \qquad (2.5)$$

where n_1 and n_0 are the refractive indices of the disperse phase and dispersion medium, respectively, and $\lambda = c_0/\nu n_0$ is the wavelength of light.

Eq. (2.5) relates to the case when the illuminating beam is linearly polarised. The intensity of light scattered in various directions is shown in Fig. 6, where x is the direction of illumination, z the direction of dipole vibration, and $\vec{r}$ the direction of irradiation.

When the illuminating light is not polarised, then the scattering diagram is different, *i.e.* it is a sum of two diagrams, in accordance with two mutually perpendicular polarised components by which non-polarised light can be represented. The corresponding formula is

$$I = \frac{9\pi^2(1 + \cos^2\theta)}{2r^2\lambda^4}\left(\frac{n_1^2 - n_0^2}{n_1^2 + 2n_0^2}\right)^2 v^2 I_0, \qquad (2.6)$$

where θ is the angle between the direction of the incident beam and the direction of observation. The scattering diagram is shown in Fig. 7. It can be seen that in this case the dipole gives rise to maximum scattering in the direction (and against the direction) of the illumination, and to minimum scattering at right angles to this direction (z), but in this case the minimum scattering is not equal to zero.

The entire light scattered by the particle, which determines the weakening of the transmitted beam (the extinction) in the case of non-polarised illumination, is obtained by integrating (2.6) with respect to θ, *i.e.*

$$R = \frac{24\pi^3}{\lambda^4}\left(\frac{n_1^2 - n_0^2}{n_1^2 + 2n_0^2}\right)^2 v^2 I_0 \tag{2.7}$$

The total amount of light scattered in a unit volume of a disperse system containing C particles, and the intensity of scattered light in a given direction are obtained for sufficiently low concentrations, as mentioned previously, by multiplying I and R, respectively, by C.

The polarisation of scattered light depends on the angle θ. The factor $(1 + \cos^2\theta)$ corresponds to two components: one is the component with a polarisation direction perpendicular to the plane formed by the incident and the scattered beam, whereas the other, which is proportional to $\cos^2\theta$, is the component polarised at right angles to the former. The scattered light is therefore non-polarised along x, and completely polarised along z, In other directions it is partly polarised.

The theory was developed further by generalising it for larger spherical particles and allowing for light absorption. This theory was developed by Love[4] in 1899 and by Mie[5] in 1908; like all other generalisations it is highly complex, and its results require the tabulation of special functions. We propose to dwell briefly on some special features of its results during the analysis of experimental data.

It has already been mentioned that scattering of light by anisodiametric particles depends on their orientation. The most complete solution of this highly complex problem was given by F. Möglich (1927)[6]. These results are so complex that they have not so far been utilised. R. Gans (1925)[7] developed the theory for small non-spherical particles. The case of cylindrical particles is here particularly simple, and can be used as an example for demonstrating the characteristics of the effect of anisodiametric form on light-scattering. When using van de Hulst's[8] formulae for the case when $kl < 1.4$ (where k is $2\pi/\lambda$ and l the length of the particle) and when the particles are very thin, we obtain

$$I_{\text{cylinder}} = I_{\text{Rayleigh}}\left[1 - \frac{(kl)^2}{18} + \frac{(kl)^4}{720}\right] \tag{2.8}$$

References p. 35

In this formula I_{Rayleigh} is identical with eqn. (2.6), whereas I_{cylinder} relates to the scattering at right angles to the direction of illumination from the unorientated particle (random orientation).

For a cylinder orientated at an angle ψ to the plane formed by illumination and observation directions we have

$$I_{\text{cylinder}} = I_{\text{Rayleigh}} \left[1 - \frac{(kl\cos^2\psi)^2}{12} + \frac{(kl\cos^2\psi)^4}{384} \right] \tag{2.9}$$

A complete treatment of the problem of light-scattering by particles can be found in the monographs of Shifrin[9] and van de Hulst[8].

(b) Scattering by fluctuations

Particularly important in the physical chemistry of macromolecules is the scattering by concentration fluctuations, which, in the case of multi-component systems, is usually considerably greater than that by the fluctuations of density. We shall therefore refrain from analysing the latter, although it is of a great fundamental significance in molecular physics.

In accordance with the theory of fluctuations* the mean-square fluctuation of concentration is given by the expression

$$\overline{\Delta C^2} = \frac{kTC}{v_N \dfrac{\mathrm{d}P}{\mathrm{d}C}}, \tag{2.10}$$

where k is Boltzmann's constant, T the absolute temperature, v_N the volume of a micro-region (fluctuation), P the osmotic pressure of the solution, and C its concentration.

When the concentration changes by ΔC, the refractive index of the solution changes by Δn, and since

$$\frac{\Delta n}{\Delta C} = \frac{\mathrm{d}n}{\mathrm{d}C} \tag{2.11}$$

we obtain

$$\overline{\Delta n^2} = \left(\frac{\mathrm{d}n}{\mathrm{d}C}\right)^2 \frac{kTC}{v_N \dfrac{\mathrm{d}P}{\mathrm{d}C}} \tag{2.12}$$

Assuming that the fluctuations are spherical (this is the most probable

* See below, the theory of Brownian movement.

assumption since the deformation of fluctuations from a sphere would require additional work), the light-scattering can be obtained from Rayleigh's equation (2.5), *i.e.* for a linearly polarised beam, allowing for (2.12):

$$I = \frac{4\pi^2 \sin^2\varphi}{\lambda^4 r^2 n_0^2} \frac{(\mathrm{d}n/\mathrm{d}C)^2}{\mathrm{d}P/\mathrm{d}C} kTCv_N I_0 \tag{2.13}$$

We utilise here the fact that Δn is very small, which results in

$$\left(\frac{n_1^2 - n_0^2}{n_1^2 + 2n_0^2}\right)^2 \approx \frac{4}{9}\left(\frac{n_1 - n_0}{n_0}\right)^2 = \frac{4}{9}\frac{(\Delta n)^2}{n_0^2}$$

and assume that $\Delta n^2 = (\Delta n)^2$.

According to (2.6) we have for the case of non-polarised illumination

$$I = \frac{2\pi^2(1 + \cos^2\theta)}{\lambda^4 r^2 n_0^2} \frac{(\mathrm{d}n/\mathrm{d}C)^2}{\mathrm{d}P/\mathrm{d}C} kTCv_N I_0, \tag{2.14}$$

and the total amount of non-polarised light scattered by a particle according to (2.7), which determines the extinction, is

$$R = \frac{32\pi^3}{3\lambda^4 n_0^2} \frac{(\mathrm{d}n/\mathrm{d}C)^2}{\mathrm{d}P/\mathrm{d}C} kTCv_N I_0 \tag{2.15}$$

When multiplied by $1/v_N$, eqns. (2.13), (2.14) and (2.15) give the amount of light scattered by 1 cm^3 of the disperse system since $1/v_N = N$ is the number of micro-volumes in 1 cm^3. Next, in these formulae $\mathrm{d}P/\mathrm{d}C$ is expressed by the dependence of osmotic pressure on the concentration by weight, C_g.

Since formula (2.10) is valid in the case of ideal solutions, as will be shown in its derivation, we must also use here a corresponding formula*,

$$P = \frac{N_A kT}{M} C_g, \tag{2.16}$$

where N_A is the Avogadro number, so that

$$\frac{\mathrm{d}P}{\mathrm{d}C} = \frac{N_A kT}{M}. \tag{2.17}$$

* *Cf.* Chapter III, Section 4, p. 55.

After substitution in, say, (2.15) and multiplication by $1/v_N$, we obtain for the total amount of light scattered by 1 cm^3 of solution*

$$R_1 = \frac{32\pi^3}{3\lambda^4 N_A n_0^2}\left(\frac{dn}{dC_g}\right)^2 C_g M I_0 \tag{2.18}$$

Since this formula is valid for an ideal solution it should be applied to data extrapolated to $C_g=0$, in the case of particle interaction, in order to determine the molecular weight M**. Analogous possibilities are also obtained from eqns. (2.13) and (2.14).

The relationship between the theory of scattering by particles and the theory of scattering by fluctuations is highly complex, and it is not possible to carry out its analysis at the level of this publication. It should be stressed, however, that although different kinds of approach were used in both theories, these theories are essentially identical, their relation being as that of particular to general, it being not always clear which of the two theories is the more general. The point is that, from the theoretical standpoint, there exists no region of conditions in which one theory is applicable and the other is not. If, *e.g.*, the conditions are such that an individual particle does not scatter light (*e.g.*, $n_1=n_0$), then the fluctuation of the particle concentration will also fail to produce light-scattering. If on the other hand the scattering particles are always distributed in the space in the same manner (*e.g.*, a uniform distribution of particles, frozen at $T=0$°K), then, from the point of view of optics (since $n_1 \neq n_0$), each particle will play the part of a fluctuation. At present the choice between these two theories being applied to a particular case depends on practical considerations. Thus, *e.g.*, when studying solutions of macromolecules, it is easy to measure n as a function of the solution concentration C; on the other hand n_1 is unknown, and we therefore use the fluctuation theory. The equivalent nature of both theories can be demonstrated as follows[1].

Note

* Owing to the structure of (2.15) this equation does not depend on the units in which concentration is expressed, and therefore we assume that $C=C_g$ when substituting.

** Usually [*e.g.* in Refs. 1 and 12] dP/dC is substituted into (2.15), calculated from the equation of state of a non-ideal solution; this is incorrect since the initial equation (2.10) is only valid for an ideal solution. This error does not entail an incorrect determination of M, since this determination is achieved by extrapolation to infinite dilution. The equation cannot, however, be used to determine the virial coefficients. For this purpose (2.10) should be replaced with a formula for an ideal solution.

Further details of molecular weight determination follow in section (*c*).

According to Wiener[10], the refractive index n of a disperse system (a fraction of whose volume, vC, is occupied by spherical particles of a volume v and refractive index n_1) is given by

$$\frac{n^2}{n_0^2} = 1 + 3\frac{n_1^2 - n_0^2}{n_1^2 + 2n_0^2} Cv$$

whence

$$\left(\frac{\mathrm{d}n}{\mathrm{d}C}\right)^2 = \frac{9n_0^2\left(\dfrac{n_1^2 - n_0^2}{n_1^2 + 2n_0^2}\right)^2 v^2}{4\left(1 + 3\dfrac{n_1^2 - n_0^2}{n_1^2 + 2n_0^2} Cv\right)} \approx \frac{9n_0^2}{4}\left(\frac{n_1^2 - n_0^2}{n_1^2 + 2n_0^2}\right)^2 v^2$$

since $Cv \ll 1$, and $(n_1^2 - n_0^2)/(n_1^2 + 2n_0^2)$ is of the order of unity. Besides for dilute solutions $\mathrm{d}P/\mathrm{d}C = kT$. By substituting this value into

$$I_1 = \frac{4\pi^2 \sin^2\varphi}{\lambda^4 r^2 n_0^2} \frac{(\mathrm{d}n/\mathrm{d}C)^2}{\mathrm{d}P/\mathrm{d}C} kTCI_0,$$

expressing the intensity of light scattered by fluctuations in 1 cm³ of the solution, we obtain the formula

$$I_1 = \frac{9\pi^2 \sin^2\varphi}{\lambda^4 r^2} \left(\frac{n_1^2 - n_0^2}{n_1^2 + 2n_0^2}\right)^2 v^2 CI_0$$

for the intensity of light scattered by particles contained in 1 cm³ of the solution. This demonstrates the equivalence of the two approaches because this equation is identical with (2.5) except for the factor C.

(c) *Experimental studies and applications*

The opalescence of colloidal solutions is studied experimentally either by measuring the intensity of scattered light at a certain angle to the direction of illumination, or by measuring the attenuation of transmitted light. The former method is often called nephelometry, the corresponding apparatus being a nephelometer, whereas the latter is applied with instruments like the conventional photometer. In the case of highly diluted sols of isodimensional, sufficiently small, non-conducting, colourless or faintly coloured particles, the results of measurements can be interpreted on the basis of Rayleigh's theory. The variable parameters used are the

References p. 35

wavelength of light, the angle at which the scattering is measured, the dilution (concentration) of the sol, and the polarisation of the scattered light. The intensity of the scattered and transmitted light is compared with that of the incident light visually or photometrically, and in recent years, also by means of photoelectric multipliers. The fluorescence effect is eliminated owing to the fact that, in this case, the wavelength of lateral light is always longer. Red light is therefore used for visual measurements of scattered light*. Since the brightness of scattered light depends to a large extent on the angle of observation, it is recommended to work with very narrow light beams and to carry out measurements making extensive use of diaphragms. All these requirements (unfortunately, not always complied with) render the study of the scattering of light by colloidal solutions a fairly complex task, which requires at any rate a well thought-out approach. Persons wishing to carry out work of this kind should acquaint themselves with the newest developments in apparatus. An adequate survey of relevant literature can be found, *e.g.* in Refs. 11, 12.

Since Rayleigh's theory involves a great number of simplifying assumptions, it cannot always be used as a basis for the treatment of experimental data. In the case of larger particles we cannot confine ourselves solely to dipoles—one must also allow for electric moments of higher order, for magnetic moments, and for the absorption of light (non-dielectrics). As mentioned on p. 18, such a theory has already been developed. Compared with Rayleigh's theory, its main feature consists in the scattering diagram being much more complex: the scattering along the direction of illumination is greater than in the direction towards the source of illumination, and the intensity of scattered light is not proportional to $1/\lambda^4$. As the wavelength is varied, the scattered light passes through extreme values, the corresponding λ values depending strongly on the particle size, which thus provides a convenient opportunity to determine this quantity. This theory was very well borne out by measurements carried out with monodisperse gold sols which, depending on the degree of dispersion, yield solutions of bright, pure colours when observed in white light. This is direct proof of the existence of sharp scattering maxima at definite wavelengths. When the theory was checked on other metallic sols (mercury, silver, platinum) the agreement with theory was not so

* The effect of fluorescence can be eliminated in every case if, when monochromatic light is used for illumination, the scattered light is measured through a monochromatic filter for the same wavelength.

close; this is explained by the anisodiametric form of the particles (similarly to Rayleigh, Mie considered the particles to be spherical) and by the fact that the optical properties of very small particles depend on their dimensions. However, these assumptions should be checked once more. Many authors have also carried out systematic investigations with non-metallic sols. The investigations of La Mer *et al.* (1949)[13] on monodisperse sulphur sols are of great interest. It was found, exactly as required by the more accurate theory (Fig. 8) that the scattered light yielded a series of maxima and minima with changing θ—the so-called Higher Order Tyndall Spectrum.

Table II,1, taken from van de Hulst[8] applies to water drops and can provide some idea of the applicability of the various degrees of approximation of the theory depending on the particle size. Since it is not the absolute magnitude of the particle, but the ratio between the diameter a of the particle and the light wavelength λ that is of importance as far as optical properties are concerned, the particle size is usually expressed as the dimensionless parameter $x = 2\pi\alpha/\lambda$.

In the case of randomly distributed anisodimensional particles, and if these particles are small, it is possible to evaluate their effective size from the light-scattering. In the case of large particles the problem is rather difficult because of the complex nature of the theory. The true dimensions and the form of particles can in this case be determined by two methods:

TABLE II, 1

THEORETICAL INTERPRETATION OF LIGHT SCATTERING

Interval x	*Diameter interval for red radiation,* μ	*Approximation theory*
0–0.5	0–0.1	Rayleigh scattering
0.5–1	0.1–0.2	Field of application of Mie's theory with three terms in expressions for the intensity of scattered light
1–10	0.2–2	Field of complex scattering diagrams—Mie's equations obtain additional terms related to electric and magnetic multipoles of a higher order. (The scattering diagram shown in Fig. 8 relates to such a case)

References p. 35

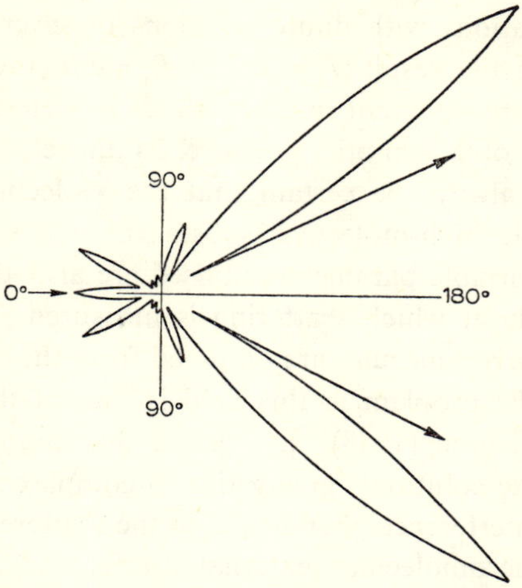

Fig. 8. Indicatrix of light scattering by a spherical particle for which $x = 2\pi a/\lambda = 6$ and $n_1/n_0 = 1.44$, calculated by Johnson and La Mer, from the Mie theory equations.

by orientating the particles and applying corresponding equations to the scattering obtained, or by utilising the phenomenon of depolarisation of the light (to evaluate the anisodiametric nature) by colloidal solutions, which had been studied in great detail by Krishnan (1935–1939)[14]. It has already been noted that, in accordance with Rayleigh's theory, the light scattered at right angles is completely polarised in a vertical plane (it is assumed that the plane illumination/observation is horizontal). However, if the particles are not spherical (anisotropic), then the direction of the induced dipoles will not coincide with the direction of the electric field, and a horizontally polarised component will appear in the light polarised at the angle of 90°, *i.e.* the light will not be completely polarised. The degree of this depolarisation is related to the anisotropy of particles and can be utilised for evaluating this anisotropy.

The studies of opalescence were developed independently for the purpose of determining the molecular weight and the form of macromolecules of high polymers (Debye, 1947[21]); the molecular weight obtained is a weight-average value. In this case one uses the fluctuation treatment of the scattering, the equations of which [*e.g.* (2.18)] include the molecular weight. The last mentioned magnitude has been introduced via the relationship between osmotic pressure and concentration. Since intermolecular interaction is negligible only at high dilution, it is necessary to carry out investigations with dilute solutions of several concentrations and to extrapolate the result ($I_1 = I/v_N$ or $R_1 = R/v_N$) to infinite dilution ($C \rightarrow 0$). This feature, inherent in all methods of determining molecular weights which involve osmotic pressure, is unwelcome in this case, since one cannot always be certain that the molecular weight of the often very unstable high-molecular substances will remain unchanged on dilution. The variable parameters utilised are also the wavelength of light and the angle at which scattering is measured. The data relating to the form of macromolecules are obtained from the degree of depolarisation. The specific problem in this field is that of the scattering of a "statistical coil" [Zimm (1948)[15]]. Flexible macromolecules can intertwine even in dilute solutions, giving rise to complex structures. This is accompanied by interference phenomena in the scattered light, caused by the intertwined macromolecules (external interference). This effect makes it necessary to introduce corrections to the theory. By comparing the theory with the experimental data it is possible to determine the mean-square distance between the ends of a flexible linear macromolecule; this distance plays an important part in the theory of polymers.

In these investigations the experimenter is faced with a number of difficult tasks. When the concentration of polymer is low, the intensity of scattered light is very weak. Any errors related to the illumination background arising from the light reflected by the walls of the measurement cell and by other parts of the apparatus, from the light scattered by dust particles in the solution and by the dispersion medium, are therefore of substantial importance. As may be seen from the formulae of scattering by fluctuations, it is necessary in such cases to measure the dependence of the refractive index of the solution on its concentration, and these measurements should be carried out with utmost accuracy.

In spite of these and a number of other difficulties, the method of studying high polymers (and the high-molecular biological systems) on the basis of light-scattering is being applied more and more widely and is undoubtedly very promising. A condensed exposition of this method (with bibliography) can be found in Refs. 11 and 12. The author of Ref. 12 believes that at present this method will enable molecular weights to be determined with an accuracy of 10–20%, although some substances yield much more accurate results.

References p. 35

2. Optical Effects due to the Orientation of Particles

It has already been noted that the brightness of light scattered by an aniso-diametric particle depends to a large extent on its orientation. This can be observed especially in the case of rod-shaped particles and, to a slightly lesser degree, in the case of lamellar particles. Thus, *e.g.*, a rod-shaped particle oriented at right angles to the illumination/observation plane scatters much more light than would the same particle if it were not orientated (random rotation). On the other hand, the light scattered by a similar particle oriented in a direction parallel to the direction of observation is much weaker than in the case of an unoriented state [*cf.* (2.8) and (2.9)].

When the particles are oriented, there arises an ordered quasi-structure which resembles a crystalline structure. Even if the particles (whose refractive index is certainly different from that of the medium) are not in themselves optically active, the entire system becomes anisotropic, and the transmitted light beam exhibits double refraction. If the particle substance is also endowed with optical anisotropy the corresponding effect is added.

It follows from the above that the optical properties of the disperse system change when the anisodiametric particles are oriented. These effects are so considerable as to be often capable of being observed without any special apparatus; it is then possible to estimate particle shape in a qualitative manner. In this respect methods based on orientation differ favourably from the extremely complicated and meticulous investigations of the anisodimensional form, based on light-scattering in the absence of orientation.

Another important advantage of the orientation methods consists in the possibility of applying or eliminating, in some cases (electric, magnetic orientation), the orienting effect at a faster rate than that at which the orientation or disorientation of particles takes place. The time of disorientation, due to Brownian motion, which is strictly a function of temperature, also permits the determination of particle dimensions for a given shape.

In addition to the purely optical aspects the theory of the phenomena considered deals also with the corresponding effect of orientation. Depending on the effect used, this part of theory makes it possible to determine additionally important properties of the particles, related also to their anisotropy with respect to the effect involved, since orientation is impossible without such anisotropy.

In view of all this, optical methods involving orientation become extremely important and highly promising for the investigation of colloidal systems.

(a) *Orientation in flow*

Taken in conjunction with birefringence measurements this orientation method is at present the most developed. The flow of colloidal solutions, and the directional motion of colloidal particles in the solution (if these particles are anisodimensional) leads to their orientation. On the other hand, Brownian movement tends to disorientate the particles at all times. If the orienting velocity gradients are not great, there arises partial orientation which increases as the orientating effect is strengthened, and tends to complete orientation when the effect is sufficiently large.

Orientation in flow is usually achieved by placing the colloidal solution in the gap between two concentric cylinders one of which is stationary, while the other rotates with a given velocity which is sufficiently small to produce laminar flow. The constant velocity gradient arising in this case is

$$q = \frac{\omega r_1}{r_2 - r_1}, \tag{2.19}$$

where r_1 and r_2 are the radii of the inner and outer cylinder, respectively, and ω is the angular velocity of rotation. The corresponding (and constant throughout the liquid) shear stress $q\eta$ (where η is the viscosity) determines the force of the orienting action.

The birefringence in the liquid layer between the cylinders is measured by conventional methods, using a light beam which passes through the layer parallel to the cylinder axis through windows in the bottom and lid of the apparatus. From the results obtained one determines the birefringence

$$M = \frac{n_1 - n_2}{n_0 c_0 q\eta} = \frac{\lambda \Delta}{n_0 c_0 q\eta l}, \tag{2.20}$$

where n_2 and n_1 are refractive indices of the ordinary (polarised perpendicular to the main axis) and extraordinary (polarised parallel to the main axis, where the latter is directed along the orienting field) beam respectively, n_0 is the refractive index of the medium, λ is the light wavelength,

References p. 35

c_0 the velocity of light in vacuum, l the length of path in the layer, and Δ the phase difference between the components 1 and 2. The theory of this effect has been developed in detail by Peterlin and Stuart[16] for particles which are small compared with λ, *i.e.* when only the induced dipole moment is allowed for.

The hydrodynamic aspect of the problem, *i.e.* the direction of orientation of the particle as a function of the direction of flow is more complex than it would appear at first sight. Orientation does not occur along the direction of flow, but at a certain angle to this direction, this angle depending also on the particle size. This peculiarity is associated with the existence of two preferential directions of orientation and with the ensuing precession of particles in the flow.

The orientating forces arising in the flow can also bring about the deformation of particles, thus producing birefringence even in some sols consisting of isodimensional (at rest) particles.

The method for studying birefringence in flow is successfully applied in the study of solutions of high polymers. It involves the investigation of the dependence of Δ (and of the associated angle of preferential orientation of particles) on the velocity of rotation, which determines the magnitude of the orienting effect. Since the orientation occurs on the background of the continuous disorientating Brownian movement, Δ increases with increasing velocity of rotation. The length of macromolecules can be estimated from such results.

The first observations of the orientation of colloidal particles in flow were made by Siedentopf (1912)[17], not on the basis of birefringence observations but from the variation in the intensity distribution of scattered light in various directions. This method provides diverse possibilities of determining the properties of colloidal particles. Its quantitative development has been undertaken recently by Nakagaki and Heller (1959)[18].

(b) *Electric orientation*

It has already been noted that the essential advantage of the orientation of colloidal particles by means of an electric and magnetic field, compared with the orientation in a flow, consists in that the orienting effect can be applied or removed virtually instantaneously. This makes it possible to investigate not only the stationary states of orientation, but also the transition states, particularly the spontaneous disorientation of particles resulting from Brownian motion. For a given particle shape this is unambiguously related to the particle dimensions which can thus be determined. Using this method, Benoit (1951)[19], and, independently, O'Konski and Zimm (1950)[20] succeeded in determining from the relaxation of disorientation (in the case of orientation induced by an electric field) the length of the tobacco mosaic virus, in good agreement with the results of electron-microscopic investigations. The main disadvantage of these methods lies in the fact that the particles must be endowed with a certain susceptibility to the action of the electric or magnetic field, but this property is by no means a universal one. The calculations carried out by the author for ellipsoids of revolution showed that diamagnetic particles have a very low susceptibility to the action of the magnetic field. It was found that, even in the strongest available magnetic fields, of the order of 100,000 gauss, the orientation energy of diamagnetic particles is comparable with the mean kinetic energy at room temperature in the case of large particles which are discernible under a microscope. Magnetic orientation cannot therefore be applied in the case of colloids, high polymers and biochemical objects, except in the special and rare cases when the substance of the particles is paramagnetic, or by a special method of magnetic induction.

The situation is much more favourable in the case of electric orientation. Investigations have shown that the electric dipole moments of anisodimensional colloid particles in aqueous solutions usually suffice for their orientation in fields in which (given a sufficient purification from electrolytes) the heating of solution by the passing electric current is still insignificant within the time during which the stationary orientation is established [Tolstoĭ (1955)[22]]. The colloidal particles and macromolecules may have their own dipole moment which is determined by their structure, and the dipole moment induced by the orientating field. When a constant electric field (or d.c. voltage pulses) is used, the orientation is due to the interaction of the field with the dipole moments of both kinds, and it is impossible to resolve them. Quite recently, the author and coworkers (1959)[23] succeeded in orienting colloidal particles (of halloysite, benzopurpurine and many others, in water) with the aid of a high-frequency electric field of the order of tens and hundreds of kc/s. It was demonstrated that the effect of the proper non-induced dipole moment, rigidly combined with the particle and therefore causing the vibration of the particle in the alternating field, is fully damped by its inertia. In this case the particles are

References p. 35

only oriented due to the induced moment, which rotates along with the field and creates a constant orientating moment of force. It was found that in aqueous solutions this moment is sufficient to orient the particles. It is probably generated owing to the surface layer on the water. If this supposition is borne out, then the electric method of orientation will prove to be a universal one for aqueous solutions. The method is also of great interest because of the possibility of using high-frequency electric fields, in which the secondary phenomena of electrolysis, polarisation and electrophoresis are considerably reduced or eliminated. By carrying out parallel investigations with a constant field it is also possible to evaluate the constant dipole moment of particles, and the angle between these two moments. Thus, it was found, *e.g.* for the particles of halloysite, that the induced moment is arranged along the threadlike particles (which agrees with the idea of its "surface" origin), whereas the non-induced moment was usually found to be directed across the particle.

The optical verification of orientation in the cases analysed in this section was effected in transmitted light by measuring the intensity variations due to orientation, or by measuring the arising birefringence. These optical methods are highly sensitive, but unfortunately the optical nature of the phenomenon involved is not always clear. The point is that the intensity variations of transmitted light during the orientation of particles can be caused by the changes of scattering and absorption, and by polarisation effects.

The optical interpretation of the phenomenon is much clearer if it is carried out on the basis of intensity variation of scattered light in a given direction. This was the method used by the author in the investigations described above, and a corresponding theory was developed for this purpose (1962)[24]; owing to its complex nature the theory will not be presented here.

The orientation effects can be approximately estimated by the following simple method. Let us substitute for the true distribution by orientation of N rod-shaped particles a distribution in which the particles can only be parallel to one of the three (x, y, z) axes of the orthogonal coordinate system. Completely random distribution will then correspond to a state in which $\frac{1}{3}$ of all particles will lie in each of the three selected directions $N_x = N_y = N_z = N/3$. If *e.g.* we have a complete orientation along x, then $N_x = N$, $N_y = N_z = 0$. The intermediate states, when the force of the electric field does not suffice to cause complete orientation, corresponds to $N > N_x > N/3$ and $N_y = N_z < N/3$, *i.e.*, as can be easily calculated from Boltzmann's law

$$N_x = \frac{N}{1 + 2\mathrm{e}^{-u/kT}} \quad \text{and} \quad N_y = N_z = \frac{N - N_x}{2}, \tag{2.21}$$

where u is the orientation energy of particles in an electric field of strength E. Let us further assume that each of the three orientations corresponds to the light intensities i_x, i_y and i_z, scattered by one particle towards the observer, and that the solution is diluted so that the total amount of scattered light is proportional to the number of particles. It is then an easy task to calculate the specific effect of variation of the intensity of scattered light in a given direction:

$$\alpha = \frac{I_E - I_0}{I_0}, \tag{2.22}$$

where I_E is the intensity of scattered light in the case of the field E, and I_0 is the intensity of scattered light in the absence of orientation. We have

$$\alpha = \alpha_\infty \frac{1 - \mathrm{e}^{-u/kT}}{1 + 2\mathrm{e}^{-u/kT}}, \quad \alpha_\infty = \frac{2i_x - i_y - i_z}{i_x + i_y + i_z}, \tag{2.23}$$

where a_∞ corresponds to complete orientation ($u \gg kT$).

The magnitude u is easily calculated for the case when the induced dipole moment attains saturation at low field values, after which it remains constant ($P = P_0 = \text{const}$), in the form

$$u = P_0 E, \tag{2.24}$$

so that

$$\alpha = \alpha(E) = \alpha_\infty \frac{1 - \mathrm{e}^{-P_0E/kT}}{1 + 2\mathrm{e}^{-P_0E/kT}}. \tag{2.25}$$

The magnitude α_∞ can be calculated from formulae describing the scattering of light by rod-shaped particles, which enables the particle length to be determined from α_∞.

The approximate nature of this calculation is characterised by the fact that the scattering of light from rod-shaped particles is very asymmetric.

3. The Ultramicroscope

Colloidal particles cannot be observed under a conventional microscope because their dimensions are smaller than the resolving power of the instrument. It is known that the resolving power is the least distance between two points which can still be perceived separately. According to Helmholtz and Abbe, it is expressed by

$$d = \frac{\lambda}{2n \sin(\alpha/2)},$$

where λ is the wavelength of the light used, n the refractive index of the medium in which the microscope objective is placed and α the aperture angle, *i.e.* the angle formed by the boundary beams proceeding from the object into the objective. When using an oil immersion for the middle part of the visible spectrum it is possible to attain a resolving power of $\sim 0.2\mu$. With ultraviolet light, special quartz optics and photographic recording, the resolving power may be increased to $0.1\ \mu$; this is the upper limit of colloidal dispersion.

A so-called *ultramicroscopic* method was suggested by Siedentopf and Zsigmondy in 1903[25]. These authors used a strong side illumination of the solution observed under a microscope, so as to prevent the light from falling into the microscope objective, *i.e.* they worked with a dark field. The colloidal particles could be seen as separate luminous points. It was therefore possible to establish their presence in the solution and to observe their movements.

The indispensable side illumination in the slit microscope of Siedentopf and Zsigmondy is achieved as shown in Fig. 9. A beam of intense light, obtained from the arc A is directed by a special optical system L_1, L_2 to the cell containing the solution under test. The slit S enables a narrow illuminated zone to be obtained in the place where beams are focused, which is indispensable when concentrated colloidal sols are being ob-

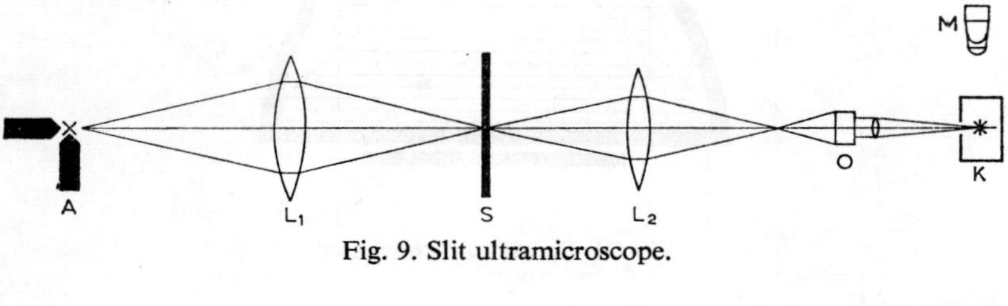

Fig. 9. Slit ultramicroscope.

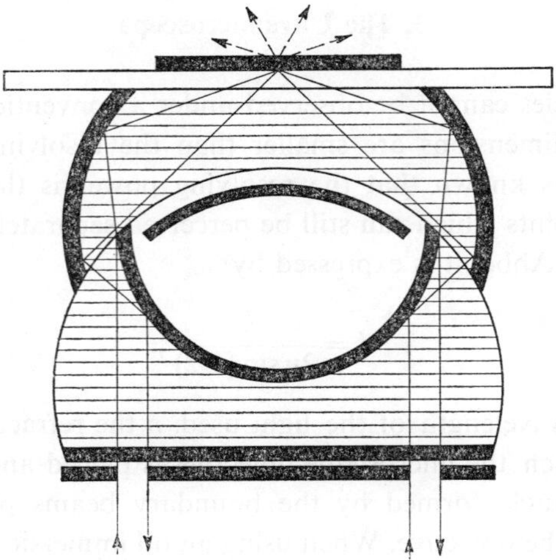

Fig. 10. Cardioid condenser.

served. Apart from this method, it is also possible to use light condensers of special design, such as the *cardioid* condenser (Fig. 10) and the *paraboloid* condenser (Fig. 11) which are substituted for the condenser of a conventional microscope. These condensers also enable the colloidal particles to be so illuminated as to prevent the light from falling directly into the microscope objective.

It is easy to see that this method is based on the microscopic observation of the Tyndall effect. Since the diffraction image round a particle is considerably greater than the particle itself, the resolving power of an ultramicroscope is higher than that of a conventional microscope. For metallic particles it attains 0.005 μ, being thus 2 orders of magnitude

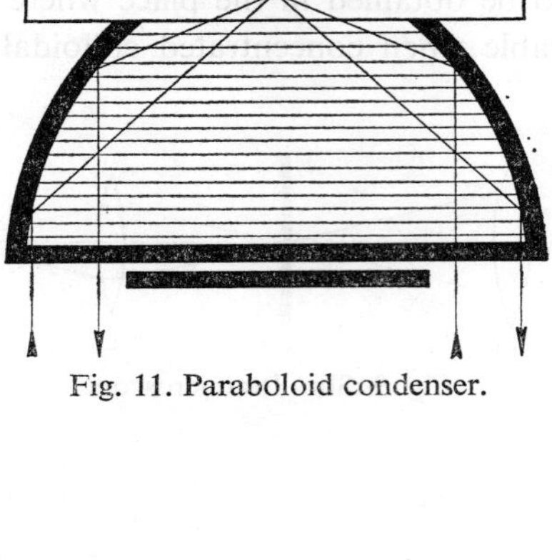

Fig. 11. Paraboloid condenser.

References p. 35

greater than in the conventional microscope. For non-metallic particles, however, the resolving power does not exceed 0.02 μ, and the method is thus unfortunately inapplicable to the study of most colloidal solutions, high polymers and biochemical objects. However, the role of this method is not confined to its historical importance (Siedentopf and Zsigmondy detected particles in a number of colloidal solutions which were empty under a microscope, and thereby confirmed the dispersion theory). At present this method is used widely for aerosols, whose particles are usually rather coarse, especially in the so-called flow ultramicroscope of Deryagin–Vlasenko[26].

REFERENCES

1 JONKER in KRUYT (Editor), *Colloid science*, Elsevier, Amsterdam, 1952.
2 TYNDALL, *Proc. Roy. Soc. (London), Ser. A*, 17 (1868) 223; Phil. Mag., 37 (1868) 394.
3 RAYLEIGH, *Phil. Mag.*, 41 (1871) 107, 274, 447; 47 (1899) 375.
4 LOVE, *Proc. Lond. Math. Soc.*, 30 (1899) 308.
5 MIE, *Ann. Physik*, 25 (1908) 377.
6 MÖGLICH, *Ann. Physik*, 83 (1927) 609.
7 GANS, *Ann. Physik*, 76 (1925) 29.
8 VAN DE HULST, *Light scattering by small particles*, Wiley, New York, 1957.
9 SHIFRIN, *Light scattering in turbid media*, *G.I.T.T.L.*, Moscow, 1951.
10 WIENER, *Koll. Beihefte*, 23 (1927) 189.
11 STACEY, *Light-scattering in physical chemistry*, Butterworths, London, 1956.
12 CH'IEN, Opredelenie molekulyarnykh vesov polimerov (*Determination of the molecular weights of polymers*), Izdatel'stro inostrannoi literatury, Moscow, 1962.
13 BARNES, KENYON, ZAISER AND LA MER, *J. Colloid Sci.*, 4 (1949) 163.
14 KRISHNAN, *Proc. Indian Acad. Sci., Sect. A*, 1 (1935) 717, 782; 5 (1937) 94, 305, 407; 7 (1938) 21, 98; 10 (1939) 395.
15 ZIMM, *J. Chem. Phys.*, 16 (1948) 1093, 1099.
16 PETERLIN AND STUART, *Z. Phys.*, 112 (1939) 1, 129.
17 SIEDENTOPF, *Z. Wiss. Mikroskopie*, 29 (1912) 1.
18 NAKAGAKI AND HELLER, *J. Polymer Sci.*, 38 (1959) 117.
19 BENOIT, *Ann. Phys. (Paris)*, 6 (1951) 561.
20 O'KONSKI AND ZIMM, *Science*, 111 (1950) 113.
21 DEBYE, *J. Phys. and Colloid Chem.*, 51 (1947) 18.
22 TOLSTOÏ, *Dokl. Akad. Nauk SSSR*, 100 (1955) 893.
23 SHELUDKO, *Annuaire Univ. Sofia*, 54 (1959/60) 229.
24 SHELUDKO AND STOILOV, *Izv. Inst. Fiziko-khim. Bulgar. Akad. Nauk*, 2 (1962) 191.
25 SIEDENTOPF AND ZSIGMONDY, *Ann. Physik*, 10 (1903) 1.
26 DERYAGIN AND VLASENKO, *Dokl. Akad. Nauk SSSR*, 63 (1948) 155.

CHAPTER III

The Molecular Kinetic Properties of Colloidal Solutions

1. Diffusion

Graham was the first to demonstrate that the diffusion of colloidal particles takes place at a much lower rate than that of molecules in true solutions. It was later established that this characteristic feature was due to the large size of colloidal particles as compared with the dimensions of molecules. Thus, the measurement of the diffusion coefficients of lyosols became one of the fundamental methods used in colloidal chemistry for the determination of the size of particles constituting the disperse phase.

In the absence of external force fields and convection, the amount of substance dm transferred across an area S during a time dt as a result of a concentration gradient dc/dx, is given by

$$dm = -DS\frac{dc}{dx}dt, \tag{3.1}$$

where D is the diffusion coefficient. Since diffusion always proceeds in the direction of decreasing concentration, the sign of dm should be opposite to that of the gradient dc/dx.

This law was formulated by Fick (1855)[1] by analogy with the law of heat transfer. Once it has been demonstrated that both processes result from the random motion of molecules, the analogy between mass transfer and heat transfer becomes quite natural. Fick's law can be presented in a simplified form if we introduce the idea of a diffusion flux, which is defined as the amount of substance transferred in a unit of time across a unit area. In accordance with (3.1) the diffusion flux can be expressed as

$$i = \frac{dm}{S\,dt} = -D\frac{dc}{dx} \tag{3.2}$$

References p. 74–75

In the most general case the diffusion flux is a function of position and time, since dc/dx depends on x and on t. If two solutions of different concentrations are brought into contact with one another diffusion will take place, tending to restore uniformity of concentration within the system. Consequently the concentration gradient will decrease, attaining zero value at equilibrium. It is only in special cases, *e.g.* when a constant concentration difference is maintained artificially in certain parts of the system that the concentration gradient—and therefore the diffusion flux—does not change with time; a steady state is established in the system. However, if the steady-state condition is not established in the system, then the flux i is a function of x and t and, correspondingly, concentration c is a function of position (x) and time (t). The relationship between c, x and t can be found as follows. Let two equal sections S be disposed at x_1 and x_2 at right angles to the direction of diffusion. According to Fick's first law, the amount

$$dm_1 = -DS\left(\frac{\partial c}{\partial x}\right)_{x=x_1} dt$$

will enter across the section at x_1 during the time dt [since $c = f(x, t)$, the derivatives with respect to x and t are partial derivatives], whereas the amount

$$dm_2 = -DS\left(\frac{\partial c}{\partial x}\right)_{x=x_2} dt$$

will leave across the section at x_2. During the time $\Delta t = t_2 - t_1$ the volume increase of the substance between x_1 and x_2 will be

$$\Delta m = \int_{t_1}^{t_2} (dm_1 - dm_2) = DS\int_{t_1}^{t_2}\left[\left(\frac{\partial c}{\partial x}\right)_{x=x_2} - \left(\frac{\partial c}{\partial x}\right)_{x=x_1}\right] dt.$$

The same increase (Δm) can be expressed by the change of concentration. In an infinitely thin layer (thickness dx) we have

$$dm = c_{t_2} S dx - c_{t_1} S dx,$$

so that

$$\Delta m = S\int_{x_1}^{x_2} (c_{t_2} - c_{t_1}) dx.$$

Equating both expressions for Δm we obtain in accordance with the theorem of mean values

$$D\left[\left(\frac{\partial c}{\partial x}\right)_{x=x_2}-\left(\frac{\partial c}{\partial x}\right)_{x=x_1}\right]_{t=t'}\Delta t=(c_{t_2}-c_{t_1})_{x=x'}\Delta x$$

where $t_1 < t' < t_2$ and $x_1 < x' < x_2$. Further, in accordance with Lagrange's finite increment theorem, we have

$$D\left(\frac{\partial^2 c}{\partial x^2}\right)_{\substack{x=x''\\t=t'}}\Delta t\Delta x=\left(\frac{\partial c}{\partial t}\right)_{\substack{x=x'\\t=t''}}\Delta x\Delta t,$$

where x'' and t'' are between x_1 and x_2, and between t_1 and t_2, respectively. On reaching the limit with respect to Δt and Δx, the values x' and x'', and t' and t'' become identical, respectively, with x and t, and we obtain:

$$\frac{\partial c}{\partial t}=D\frac{\partial^2 c}{\partial x^2} \tag{3.3}$$

This relationship is sometimes called Fick's second law. By solving this differential equation we can express concentration as a function of position and time, *i.e.* $c=c(x, t)$, if we know the boundary conditions which determine the form of the solution obtained.

Formula (3.3) can be used not only in cases when diffusion proceeds in the x direction, but also for the two-dimensional or three-dimensional diffusion. In such cases we must use instead of (3.3) a more complex expression, *viz.*

$$\frac{\partial c}{\partial t}=D\Delta c \tag{3.4}$$

where Δ is the Laplace operator; thus

$$\Delta c=\frac{\partial^2 c}{\partial x^2}+\frac{\partial^2 c}{\partial y^2}+\frac{\partial^2 c}{\partial z^2}.$$

In the laws analysed above, the diffusion coefficient D is the only magnitude which depends on the properties of the dispersed substance and of the disperse medium. We find from (3.2) the dimension of this constant [$cm^2\ sec^{-1}$]. Since diffusion is a slow process, the values of D are low. The highest D values are found in gases whose molecules are endowed with the highest mobility. In this case we obtain from the kinetic theory of gases the following expression for the diffusion coefficient.

$$D=\tfrac{1}{3}\lambda\bar{u}\ ,$$

References p. 74–75

where λ is the mean free path, and $\bar{u}$ the mean velocity of molecules. In the case of condensed systems the calculation of D is more complicated. A theory for this case was put forward by Frenkel (1926)[2]. According to this theory D increases exponentially with increasing temperature.

The diffusion coefficient of colloidal particles is related to their coefficient of friction, B, by the expression

$$D = \frac{kT}{B}.$$

This formula is derived on p. 45. For spherical particles of radius r which is large compared with the radius of the solvent molecules, and when the viscosity of the disperse medium is η, this coefficient is $6\pi\eta r$.

Fick's law enables the diffusion coefficient to be determined. To achieve this we carefully establish contact between the solution of the substance whose diffusion coefficient is to be determined and the solvent, so as to obtain as distinct a boundary as possible between these two liquids. The system is left to stand for a long time in conditions which exclude the possibility of convection-induced mixing; *i.e.* at a strictly constant temperature and in the absence of vibrations. After some time the new distribution of concentrations is determined and the diffusion coefficient is calculated from eq. (3.4).

Many methods are available for determining the attained distribution of concentration. In some cases the solution is divided into four equal parts, the boundary planes being at right angles to the direction of diffusion, and the mean concentration is determined for each of these parts; when the time from the beginning of the experiment is known, D may be found* from these concentration values with the aid of special tables calculated on the basis of eqn. (3.3). In other cases the distribution of concentrations is determined by optical methods, utilizing, *e.g.* the relationship between the refractive index of the solution and its concentration, or the dependence of other optical properties of the solution on its concentration. The apparatus suggested by Chmutov and Slonim is very convenient for this purpose.

This apparatus (Fig. 12) consists of the casing C and slide D, which are clamped between two parallel plates A and B. D can be moved by means of the screw G. Before starting the measurements the slide is shifted so as

* The determination of concentration in various layers of the solution for the purpose of determining the rate of diffusion had been undertaken by Graham.

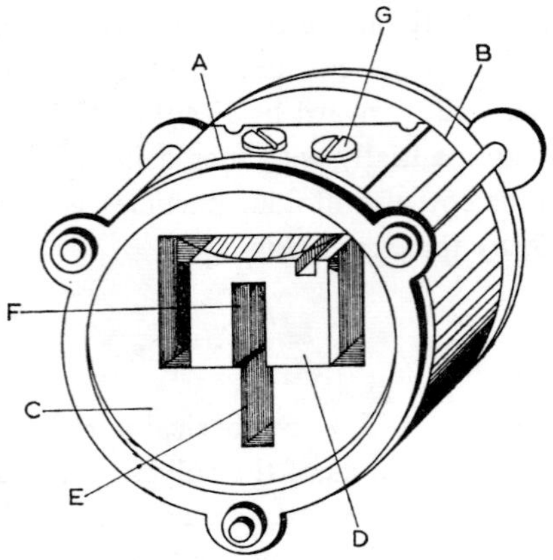

Fig. 12. The apparatus of Chmutov and Slonim for the determination of *D*.

to render the slits E and F non-coincident, after which E is filled with the solution, and F with the solvent. The slide is then moved until E coincides with F, and a sharp boundary between solution and solvent is formed. After some time the actual distribution of the diffusing substance throughout the entire volume is determined by letting light pass through the slit E–F.

The main difficulty encountered in the accurate measurement of the diffusion coefficient consists in that it is indispensable to maintain a constant temperature, to within $\pm 0.1°$, over a long period of time. Because of that, and due to many other difficulties, the accuracy of measurement of *D* rarely does exceed 1 %.

2. Osmotic Pressure

Similarly to the osmotic pressure of ordinary molecular solutions, the osmotic pressure *P* of colloidal solutions is described in the most general form by the so-called virial form

$$P = c_g RT\left(\frac{1}{M} + A_2 c_g + A_3 c_g^2 + \ldots\right), \tag{3.5}$$

where c_g is the concentration by weight, *R* the gas constant, *T* the absolute temperature, *M* the molecular weight, and A_2, $A_3 \ldots$ are virial coefficients which express the intermolecular interactions and other factors by which a given solution differs from an ideal one.

References p. 74–75

In the case of highly diluted solutions eqn. (3.5) becomes

$$P = \frac{c_g}{M} RT = c^* kT, \tag{3.6}$$

where $k = R/N_A$ is Boltzmann's constant, and c^* is the particle concentration ($c^* = c_g N_A/M$). It can be seen that the magnitude of osmotic pressure is determined in principle (ideal solution) by the number of molecules or particles in a unit volume.

Solutions of lyophobic colloids usually contain a low concentration of particles. In the case of gold sols one cannot obtain more than 10^{15} particles per 1 cm^3 of solution. According to (3.6) this corresponds, at room temperature ($kT \approx 4 \cdot 10^{-14}$), to $P = 40$ dyne/cm^2 or $4 \cdot 10^{-5}$ atm. Such insignificant osmotic pressures cannot be measured directly in an osmotic cell, or indirectly by the ebullioscopic or cryoscopic methods. Besides, when using these indirect methods one has to subject the solution to freezing or boiling, which often results in coagulation because of the instability of colloidal solutions. For these reasons the method of determining the size of colloidal particles from the measurement of osmotic pressure, using formulae (3.5) and (3.6), cannot be applied to lyophobic colloids. On the other hand, this method is widely used for the determination of the molecular weights of macromolecular solutions. The particles (macromolecules) contained in these solutions are much smaller than those in the colloidal solutions of suspensoids, and moreover the former solutions are much more stable than the latter. It is therefore possible to obtain c^* values which give rise to measurable values of P. However, the latter values are still low, particularly since for the calculation of molecular weights it is necessary, in accordance with (3.5), to carry out the measurements at low values of c_g to make the extrapolation to $c_g = 0$ reliable (where the terms containing the virial coefficients vanish). This complicates the problem, since we have to satisfy simultaneously the following main requirements:

(1) Small c_g values, and the corresponding small P values, which requires a high accuracy of measurement of the small rises of the liquid meniscus in the osmometer.

(2) Rigid construction of the osmotic cell, its small volume, and absence of gas bubbles, so as to minimize any oscillations of the liquid level above the osmometer.

(3) High degree of constancy of temperature (for the same reason).

(4) Highly reliable semi-permeable membranes, which completely retain the macromolecules, are perfectly permeable to the solvent, and are insoluble in the solvent.

(5) Furthermore, the osmometer design and the permeability of membranes should enable the osmotic pressure to be established in a short time.

All these requirements cannot easily be met. There exist nearly 50 osmometers, even the best of which, *e.g.* the osmometer of Fuoss and Mead (1943)[3], are not free from drawbacks. Many procedures for the preparation of semi-permeable membranes have been developed in recent years, mainly on the base of synthetic polymers. When the membranes are prepared by a special, and highly complex, method, it is possible to determine with their aid molecular weight as low as 2000, but the usual lower limit is approximately 30,000. Apart from the reading of the equilibrium rises of the osmometer liquid level, ingenious dynamic methods are also used, in which the osmotic rise is calculated from the rate of passage of the solvent into the osmotic cell; the time of measurement is thus considerably reduced.

All these techniques form a special branch of experimental methods, *viz.* that of the osmometry of solutions of high-molecular substances. For details the reader is referred to special textbooks, *e.g.* in refs. 4 and 5. Although all these methods have, to a large extent, good theoretical foundations, this branch of science is still being developed mainly in connection with the development of macromolecular chemistry and technology.

Macromolecular systems are usually polydisperse. Consequently the osmotic pressure of solutions is in this case the overall effect. We can write

$$P = RT\sum\frac{c_{gi}}{M_i} = RTc_g\frac{1}{\overline{M}_n}, \tag{3.7}$$

where $\overline{M}_n$ is the number average molecular weight of all the n components present; this is in fact the value determined by the osmometric method. The accuracy with which it is at present possible to determine these mean molecular weights does not apparently exceed 20 % in the most favourable cases.

It was tacitly assumed in the cases so far analysed that the colloid is not an electrolyte; this is true in all cases of macromolecular solutions in non-polar solvents. In the case of aqueous solutions and, in the first place, in the case of various bio-colloids, the macromolecular part is usually an ion. If the solution also contains electrolytes which can permeate through the membrane, the matter becomes complicated. When osmotic equilibrium is attained, it is found that the usual electrolyte is not uniformly distributed on both sides of the membrane; the resulting osmotic pressure is less than that one would obtain if there were no electrolyte. Unless we allow for this lowering, the measured molecular weights will be too high. The theory of this effect was given by Donnan in 1911[6]. Let us assume that a solution of the colloidal electrolyte NaR (concentration c_1) is on one side of the membrane, whereas a solution of an ordinary electrolyte, *e.g.* NaCl (concentration c_2)—on the other. Following Donnan, equal volumes are taken on both sides of the membrane. The membrane is perfectly permeable to the solvent (water) and the ions Na^+ and Cl^-, but is impermeable to R^-. To simplify matters, let us assume also that both electrolytes are completely dissociated. After the equilibrium has been established we shall find x moles of NaCl in the NaR compartment; therefore the concentration of Na^+ will rise to c_1+x, that of R^- will remain c_1, and that of Cl^- will be x. On the other side of the membrane the concentration of Na^+ and Cl^- will drop to c_2-x. Since equilibrium is now established, the free energy variation (the osmotic work of transporting dn moles of salt passing across the membrane) is equal to zero:

$$\underbrace{\mathrm{d}nRT\ln\frac{c_2-x}{c_1+x}}_{\mathrm{Na}^+} + \underbrace{\mathrm{d}nRT\ln\frac{c_2-x}{x}}_{\mathrm{Cl}^-} = 0$$

Hence

$$x = \frac{c_2^2}{2c_2+c_1},$$

and the measured osmotic pressure, which is equal to the difference between the osmotic pressures of both electrolytes, will be

$$P = \underbrace{2RTc_1}_{\mathrm{RNa}} - \underbrace{2RT[(c_2-x)-x]}_{\mathrm{NaCl}},$$

and, after substituting for x,

$$P = 2RT\frac{c_1^2+c_1c_2}{c_1+2c_2}$$

Consequently, the ratio of the measured osmotic pressure to the osmotic pressure of RNa will be $(c_1 + c_2)/(c_1 + 2c_2)$.

When there is an excess of NaCl ($c_2 \gg c_1$), the measured pressure will be one-half of the value one would obtain had the Donnan effect not been allowed for. Obviously this effect will play no part in the opposite case when $c_1 \gg c_2$.

3. Diffusion and Osmotic Pressure

In 1888 Nernst[7] put forward a theory of diffusion in which he assumed that the cause of diffusion is the osmotic pressure. Of course, in the case of diffusion osmotic pressure cannot be considered as a real force, since this pressure is only observed in the presence of a semi-permeable membrane. However, osmotic pressure may be interpreted as a measure of the chaotic motion of molecules, which results in an equilibrium distribution of components in the system. Such an approach to diffusion lacks so far strict theoretical foundations, but the remarkable results obtained with it justify its application.

Let us assume that a tube with a cross-section area of 1 cm^2 is filled with a solution; let us consider in this tube a layer of thickness dx, bounded by two semi-permeable membranes. Let the difference of concentration between both sides of the layer be dc^* (Fig. 13). In this case the volume $1 \cdot \mathrm{d}x$ will be exposed to a force which is equal to the difference between the osmotic pressures on both sides $\mathrm{d}P = -kT\,\mathrm{d}c^*$. In this case

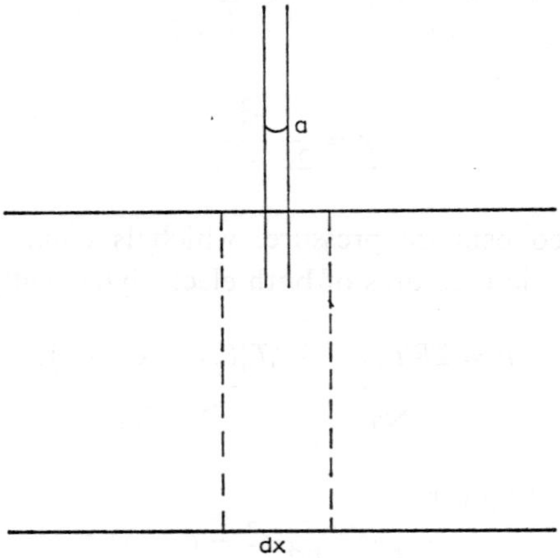

Fig. 13. Tube filled with a solution and two semi-permeable membranes.

References p. 74–75

the pressure is completely real and measurable: it can deflect the membranes, or even burst them if they are not sufficiently strong. No mass transfer will occur, as the system is in equilibrium. If we connect the volume $1 \cdot dx$ with an open manometric tube, the solvent will begin to pass through the two membranes, and liquid in the tube *a* will rise until the hydrostatic and osmotic pressures are balanced. If instead of connecting the volume $1 \cdot dx$ with a tube we remove the partitions, the diffusion process will set in, *i.e.* the migration of the dispersed substances from the region with a higher concentration to that with a lower concentration. It is natural to follow Nernst by assuming that in the latter case the same force is active as that which causes osmotic pressure in the presence of a semi-permeable membrane. The only difference is that this force acts on the particles of the dispersed phase and not on the membranes. Since c^*dx particles are present in the volume $1 \cdot dx$, the force f which acts on any one particle is

$$f = -kT\frac{dc^*}{c^*dx}.$$

It is easy to calculate from it the diffusion coefficient. If the particles acted upon by the force f move uniformly, then $f = Bu^{\dagger}$, where B is the coefficient of friction between particles and the surrounding medium, and u is their velocity. The number of particles transported in 1 sec across a unit area at x, *i.e.* the flux i_x, is uc^*_x. Thus

$$i_x = -\frac{kT}{B}\left(\frac{dc^*}{dx}\right)_x$$

By comparing this equation with (3.2) we obtain

$$D = \frac{kT}{B} \tag{3.8}$$

† In the case of a small velocity in a medium which offers a resistance (friction), the particles acquire a constant velocity. It can be demonstrated that in the case of colloidal particles the rate of diffusion is sufficiently small for this condition to be complied with, and that moreover particles acquire the constant velocity of migration almost instantaneously.

It is difficult to determine the value of B for very fine particles whose size is comparable with that of the molecules of the disperse medium, since in this case the medium cannot be considered as continuous with respect to the particle. On the other hand, in the case of colloidal solutions, where particles are large in relation to the solvent molecules, B has a simple physical significance. For instance, in the case of spherical particles we have, according to Stokes, $B = 6\pi\eta r$, provided the conditions of their motion correspond to assumptions made in the derivation of this formula, *i.e.* that the particles are spaced sufficiently far away from each other and from the walls of the vessel and that their velocities are small. These assumptions are usually valid for colloidal solutions, so that the formula

$$D = \frac{kT}{6\pi\eta r} \tag{3.9}$$

is correct for colloidal particles of spherical shape, where r is the radius of the particle, and η the viscosity of the disperse medium. For non-spherical particles, which can be considered as ellipsoids of revolution, Perrin deduced the following formula:

$$\frac{D}{D_0} = \frac{(a/b)^{\frac{2}{3}}}{\sqrt{1-(a/b)^2}} \ln \frac{1+\sqrt{1-(a/b)^2}}{a/b} \tag{3.10}$$

where D_0 is the diffusion coefficient of spherical particles of the same volume, and a and b are the minor and major semi-axes of the ellipsoid of revolution.

It may be seen from what has been said before that D is determined by the viscosity of the disperse medium, which sharply decreases with increasing temperature. For this reason, and because of the proportionality between D and kT, the magnitude of D increases rapidly with rising temperature. The magnitude of B depends on the dimensions and shape of particles. Therefore D is a constant which depends not only on the size of the particle, but also on its shape.

Finally, it is necessary to allow for the fact that B is determined for the actual dimensions of the particles during their movement in the disperse medium, *i.e.* for particles surrounded by their shells of solvation. When the solvation is strong (association with considerable quantities of the solvent) and the particles are small, this results in an increase of B as compared with the value that would be obtained by calculating B merely on the basis of the volume of the dispersed substance.

References p. 74–75

4. Brownian Movement

In 1828, during a microscopic examination of a suspension of pollen in water, Brown noticed[8] that the pollen grains were in a state of continuous motion. This phenomenon, called later the Brownian movement, could not be explained for a long period of time. The assumption that the movement of pollen grains was caused by their vital activity was rejected, since the same phenomenon was also observed in suspensions of inorganic substances. The effect could also not be explained by the formation of microscopic convection currents in the system due *e.g.* to oscillations of temperature. Had the movements of particles been brought about by convection, adjacent particles (*i.e.* particles in the same current) should have the same direction and the same velocity. In actual fact however, adjacent particles move with different velocities and their trajectories intersect each other at diverse angles. Consequently, the cause of the Brownian movement is more "microscopic" than micro-convective in nature.

Echsner (1900)[9] was the first investigator who tried to check experimentally whether this movement was of a molecular-kinetic nature. If this were the case then according to the law of equipartition of energy among the degrees of freedom, the kinetic energy of the translational movement of suspended particles should be the same as that of the translational movement of the molecules of the disperse medium, *i.e.*

$$\tfrac{1}{2}\, m\, \overline{u^2} = \tfrac{3}{2} kT\,,$$

where $\sqrt{\overline{u^2}}$ is the root mean square velocity of colloidal particles with a mass m. Echsner determined $\overline{u^2}$, by taking the mean value of particle displacements observed under the microscope over 1 sec, and obtained for the known mass m a velocity smaller by the factor of 1000 than the velocity calculated from the equation referred to above. This divergence is explained by the fact that it is impossible to observe by means of the microscope the true paths of particles undergoing Brownian movement, since owing to the limited resolving power of the microscope and to the inertia of the human eye it is impossible to discern all movements and vibrations of the particles. This results in the found velocities being considerably lower than the actual values.

In 1905 Einstein[10b], and independently Smoluchowski (1906)[10c], developed a quantitative theory of Brownian movement, which was later brilliantly confirmed by the experimental work of Perrin[11], Svedberg and others.

The theory of Brownian movement, according to which the visible movement of microscopic bodies is explained by the random thermal movement of molecules, became a direct proof of the existence of molecules, precisely at the time when this existence was fiercely discussed. The theory decided this argument in favour of the ideas of the molecular kinetic theory, and demonstrated the lack of support to the assertions of Ostwald's "energeticians" who rejected the real existence of molecules. The theoretical device used in this theory is a brilliant achievement of statistical physics. The theory of Brownian movement is the first quantitative theory of colloid chemistry.

In 1908, at the suggestion of Lorentz, Einstein published[10b] the simplified theory of Brownian movement, expounded in a more accessible form. Let us analyse this elementary theory with certain simplifications required of the sake of clarity.

Let us imagine a plane M passing through the colloidal solution (Fig. 14) and let the concentration c_1^* on the left-hand side be greater than c_2^* on the right-hand side. In both parts of the solution colloidal particles perform Brownian movement as a result of the random molecular motion. The cause of Brownian movement is the difference in the sum of impacts

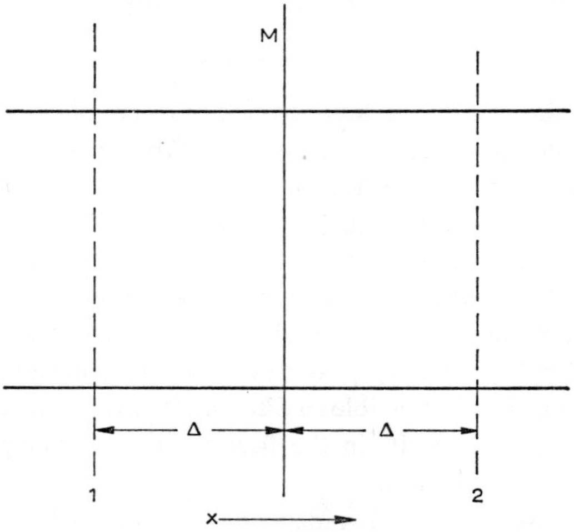

Fig. 14. Brownian movement in colloidal solution (simplified Einstein theory).

received from molecules which hit the particle from opposite sides. It is obvious that the smaller the particle the less probable it is that the molecular bombardment will be exactly balanced, and, therefore, the more intensive will be Brownian movement. At a given temperature the mean displacement Δ in the direction of the x-axis during time t will be the same to the left and to the right, for particles of the same size. The number of particles which will traverse the distance Δ from plane 1 to plane M will therefore be half the product Δc_1^*. The number of particles which traverse during the same time t, in the opposite direction, the distance Δ from plane 2 to plane M will correspondingly be $\frac{1}{2}\Delta c_2^*$. It follows that the number of particles transported across plane M in time t is

$$\tfrac{1}{2}\Delta c_1^* - \tfrac{1}{2}\Delta c_2^* = \tfrac{1}{2}\Delta(c_1^* - c_2^*).$$

If Δ is sufficiently small (in the case of a sufficiently small t), we have

$$\frac{c_1^* - c_2^*}{\Delta} = -\frac{\mathrm{d}c^*}{\mathrm{d}x}$$

(the negative sign being used for the same reason as in Fick's law). Consequently, the quantity of substance transported during 1 sec will be

$$-\frac{\Delta^2}{2t}\frac{\mathrm{d}c^*}{\mathrm{d}x}$$

If the area of plane M is 1 cm^2, we obtain the diffusion flux

$$i = -\frac{\Delta^2}{2t}\frac{\mathrm{d}c^*}{\mathrm{d}x}.$$

Comparing this expression with Fick's first law, (3.2), we obtain

$$D = \frac{\Delta^2}{2t} \quad \text{or} \quad \Delta^2 = 2Dt. \tag{3.12}$$

In deducing this formula we divide by t the quantity of substance $\frac{1}{2}\Delta^2\,\mathrm{d}c^*/\mathrm{d}x$ transported during the time t, in order to find the quantity of substance transported during a unit of time. This means that the mean displacement of the particle in any direction is assumed to be proportional to time, or, in other words, that this "mean" movement is considered to be uniform.

This approximation is inexact, since when the direction and velocity of the movement change, the particles move for some time with an acceleration. Eqn. (3.12) is therefore exact only for such values of t which are sufficiently large in comparison with the time necessary for the uniform movement to be established. A more detailed deduction of (3.12) makes it possible to evaluate the minimum time for which this equation is still valid, and to determine exactly the nature of the mean displacement Δ. This deduction, suggested by Langevin[12], is given below.

In accordance with the definition of the concept of "force" there exists, for one of the three components of translational movement, the following relation between the force acting on the particle, its velocity dx/dt and acceleration d^2x/dt^2:

$$m\frac{d^2x}{dt^2} = f - B\frac{dx}{dt}, \tag{3.13}$$

where m is the mass of the particle, f the external force, and $B(dx/dt)$ the frictional force, provided that the particle does not move very fast. This expression cannot be directly utilised for finding the mean displacement in the x-direction, since the mean values of velocity and acceleration are zero (the probabilities of the particle having identical values of the velocities dx/dt and $-dx/dt$ and of the corresponding accelerations d^2x/dt^2 and $-d^2x/dt^2$ are equal in each case). It is therefore necessary to transform the above expression so as to have in it the squares of these magnitudes. To this end both parts of eqn. (3.13) are multiplied by x, to give

$$\frac{m}{2}\frac{d^2(x^2)}{dt^2} - m\left(\frac{dx}{dt}\right)^2 = fx - \frac{B}{2}\frac{d(x^2)}{dt}$$

Averaging this expression for a large number of particles, we obtain

$$\frac{m}{2}\cdot\overline{\frac{d^2(x^2)}{dt^2}} - kT = -\frac{B}{2}\cdot\overline{\frac{d(x^2)}{dt}},$$

since $\overline{m(dx/dt)^2} = kT$. Integration of this expression gives

$$\overline{\frac{d(x^2)}{dt}} = \frac{2kT}{B} - Ke^{-Bt/m}$$

where K is an integration constant. In the case of sufficiently small t, at which $Ke^{-Bt/m}$ is small compared with $2kT/B$,

$$\frac{\overline{\mathrm{d}(x^2)}}{\mathrm{d}t} = \frac{2kT}{B}$$

which after integration and replacement of kT/B by D, according to (3.8), becomes

$$\overline{x^2} = 2Dt\,. \qquad (3.12')$$

This expression coincides with (3.12) if we assume that $\Delta^2 = \overline{x^2}$. Clearly, in Einstein's equation (3.12) we are dealing with the mean square of the displacement.

It can be seen that (3.12′) is valid for times t greater than $(m/B) \ln (K/2D)$. This magnitude is of the order of 10^{-8} sec. Thus, Einstein's formula (3.12′) is valid for a very long, virtually infinite, interval of time.

An experimental check of the theory of Brownian movement was carried out by Perrin, Svedberg and others with the aid of eqn. (3.12′), and it has brilliantly corroborated the theory. Perrin[11] observed under the microscope the movement of one particle, noting its position in the field of vision at regular intervals of time. His results are shown in Fig. 15, which represents a diagrammatic projection of Brownian movement on a plane. If we project the obtained sections on to an arbitrary straight line, we shall obtain a projection of Brownian movement in one direction. By squaring and averaging such projections, we obtain $\overline{x^2}$. A gamboge suspension, whose particles are spherical ($B = 6\pi\eta r$), was used in the experiments. The radius r of particles was determined under the microscope. Using eqn. (3.12′), where $D = kT/6\pi\eta r$, Perrin calculated the Boltzmann constant and hence Avogadro's number N_A ($k = R/N_A$, where R is the gas constant), obtaining a value of $6.5 \cdot 10^{23}$. This is in remarkable agreement with N_A

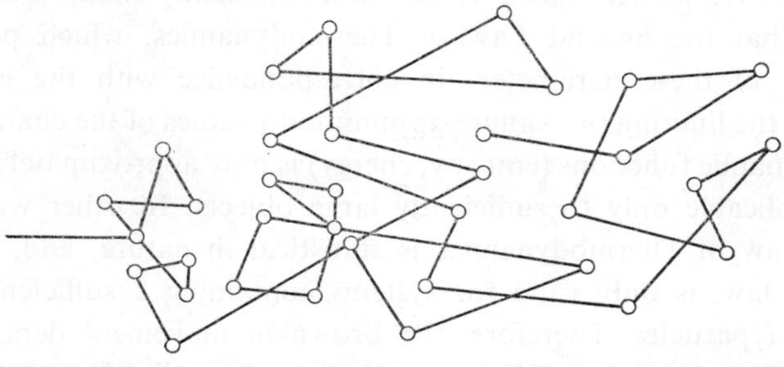

Fig. 15. Projection of Brownian movement on a plane. The position occupied by the particle at equal time intervals is indicated by small circles.

determined by other methods ($6.03 \cdot 10^{23}$), bearing out the correctness of molecular-kinetic concepts.

The theory of Brownian movement also makes it possible to calculate a number of other magnitudes which characterize this phenomenon, *e.g.* the square of the mean angular displacement of a particle during Brownian rotation, and the deviation of the number of particles in a given volume from the mean concentration, when this number is determined at regular time intervals.

For the rotation of one particle an expression is obtained which is completely analogous to (3.12′), *i.e.*

$$\overline{\varphi^2} = 2\theta t,$$

where θ is a magnitude which, similarly to the diffusion coefficient, determines the velocity of the rotational movement of the particle, caused by the random molecular bombardment, and is the ratio of the mean kinetic energy kT to the coefficient of friction B' when the particle rotates in a viscous medium; $\overline{\varphi^2}$ is the mean square of the angular rotation about a certain axis, and t is the time during which this displacement takes place. Perrin checked the validity of this formula by observing the rotation of a suspended spherical particle by the displacement of a distinguishing mark on its surface.

Brownian movement is closely related to the fluctuations of parameters characterizing the equilibrium state of a system about their mean values. Thus, *e.g.* the lack of compensation in the sum of impacts received by colloidal particles from all sides corresponds essentially to such variations of pressure. These fluctuations are the spontaneous deviations of parameter values from their mean values in a sufficiently small volume, and indicate that the Second Law of Thermodynamics, which postulates constancy of these parameters in correspondence with the condition regarding the limiting (maximum or minimum) values of the characteristic thermodynamic functions (entropy, energy) is only approximately correct, being applicable only to sufficiently large objects. In other words, the Second Law of Thermodynamics is statistical in nature, and, as every statistical law, is only valid for systems containing a sufficiently large number of particles. Therefore, the Brownian movement demonstrates *ad oculos* the correctness of Boltzmann's views regarding the probabilistic nature of the Second Law of Thermodynamics, this is the qualitative

References p. 74–75

difference between the First and Second Laws of Thermodynamics, at least in the light of present-day concepts.

The theory of fluctuations, of which Brownian movement is a special case, is based on the following general postulates. The average state of a sufficiently small region of the system is characterized by a definite value of some characteristic thermodynamic functions, *e.g.* the free energy. According to the Second Law of Thermodynamics, the free energy of a system in a state of equilibrium has a minimum value at constant temperature. If the state in the analysed region deviates slightly from the "average" state, say as a result of random causes, then the free energy will increase from F to $F+\Delta F$, as a result of variation in a certain parameter λ. In accordance with Boltzmann's principle, the probability of such change will be

$$w = \text{const.}\, e^{-\Delta F/kT} \tag{3.14}$$

For small fluctuations of λ the increment ΔF can be expanded into a series:

$$\Delta F = F'(\lambda_0)\Delta\lambda + \frac{F''(\lambda_0)(\Delta\lambda)^2}{2} + \ldots \tag{3.15}$$

Taking into account the fact that the minimum of F occurs at $\lambda = \lambda_0$ (equilibrium state), we can write

$$F'(\lambda_0) = 0 \quad \text{and} \quad F''(\lambda_0) > 0 \tag{3.16}$$

Considering only the first two terms of the series, we obtain

$$\Delta F = F''(\lambda_0)\frac{(\Delta\lambda)^2}{2}$$

and

$$w = \text{const.}\, \exp[-F''(\lambda_0)(\Delta\lambda)^2/2kT] \tag{3.17}$$

This is the Gaussian distribution function well-known from the theory of probability.

Since the sum of probabilities of all fluctuations is $\int_{-\infty}^{+\infty} \mathrm{d}w = 1$, the constant in (3.17) is:

$$\left(\int_{-\infty}^{+\infty} \exp\{-F''(\lambda_0)(\Delta\lambda)^2/2kT\}\mathrm{d}\lambda\right)^{-1}$$

Since the mean-square value of the fluctuating parameter is

$$\overline{(\Delta\lambda)^2} = \int_{-\infty}^{+\infty} (\Delta\lambda)^2 \, \mathrm{d}w$$

we obtain finally

$$\overline{(\Delta\lambda)^2} = \frac{\displaystyle\int_{-\infty}^{+\infty} (\Delta\lambda)^2 \exp\left[-\frac{F''(\lambda_0)(\Delta\lambda)^2}{2kT}\right] \mathrm{d}\lambda}{\displaystyle\int_{-\infty}^{+\infty} \exp\left[-\frac{F''(\lambda_0)(\Delta\lambda)^2}{2kT}\right] \mathrm{d}\lambda} = \frac{kT}{F''(\lambda_0)}. \quad (3.18)$$

Eq. (3.18) is very convenient and is often used for evaluating the magnitude of fluctuations.

When investigating the problem of light scattering by the fluctuations of concentrations (Chapter II) we used a formula for the mean-square fluctuation of concentration $\overline{(\Delta c)^2}$. This formula can be obtained from (3.18) by assuming that $\lambda = c$. Let us imagine that the solution volume, *e.g.* 1 cm^3, is subdivided into a large number, N, of equal small elements v_N. For the variation of free energy we use the known relationship

$$\mathrm{d}F = -S\,\mathrm{d}T - P\,\mathrm{d}v + \mu\,\mathrm{d}n,$$

where S is the entropy, P the pressure, v the volume, μ the chemical potential of the dissolved component, and n the number of dissolved molecules in volume v_N. In the present case the change of F is solely due to the variation of concentration, and therefore

$$T = \text{const.} \quad (S\,\mathrm{d}T = 0) \quad \text{and} \quad v = \text{const.} \quad (P\,\mathrm{d}v = 0).$$

By neglecting secondary effects such as the change of F of volume external with respect to v_N when dn molecules pass from the space surrounding v_N ($v_N \ll 1$ cm^3), and the contraction of the solution when the concentration is varied, we obtain

$$\mathrm{d}F = \mu\,\mathrm{d}n \quad \text{and} \quad \frac{\mathrm{d}F}{\mathrm{d}c} = \mu\frac{\mathrm{d}n}{\mathrm{d}c}.$$

Since $n = cv_N$, we obtain dn/d$c = v_N$ and since $v_N =$ const.,

$$\frac{\mathrm{d}^2F}{\mathrm{d}c^2} = v_N\frac{\mathrm{d}\mu}{\mathrm{d}c}.$$

It is also necessary to know the relationship $\mu = \mu(P)$. In the general case we can write

References p. 74–75

$$\mu = \mu_0 + kT \ln \gamma P,$$

where μ_0 is the standard value of μ, P is the osmotic pressure, and γ a coefficient which includes all factors distinguishing the given solution from an ideal one (at $\gamma = 1$, $\mu = \mu_0 + kT \ln P$; an ideal solution case). Differentiation of μ with respect to c gives

$$\frac{d\mu}{dc} = kT \left[\frac{1}{\gamma} \frac{d\gamma}{dc} + \frac{1}{P} \frac{dP}{dc} \right]$$

from which, using (3.18),

$$\overline{(\Delta c)^2} = \frac{1}{v_N} \frac{1}{\frac{1}{\gamma} \frac{d\gamma}{dc} + \frac{1}{P} \frac{dP}{dc}} \,. \tag{3.19}$$

The formula used in the evaluation of light scattering from fluctuations (see 2.10):

$$\overline{(\Delta c)^2} = \frac{1}{v_N} \frac{kTc}{dP/dc} \tag{3.19'}$$

is only obtained in ideal solutions, for which $\gamma = 1$, $d\gamma/dc = 0$ and $P = kTc$.

5. Sedimentation

In the analysis of the phenomenon of diffusion we have so far neglected the effects of external force fields on the system, assuming that the components are uniformly distributed in the state of equilibrium. This is, however, incorrect in the case of sufficiently heavy particles since such particles are markedly influenced by the terrestrial gravitational field and owing to their weight, settle at an appreciable rate. As a result of this sedimentation the uniform distribution is disturbed, and conditions arise which favour diffusion in the opposite direction; an equilibrium is finally established at which the distribution of the dispersed substance is no longer uniform.

To simplify the problem let us first analyse sedimentation without taking diffusion into account.

The force of gravity which causes the sedimentation can be expressed as $f = mg$, where m is the effective mass of the particle, equal to its mass in vacuum less the mass of the displaced disperse medium, and g is the acceleration due to gravity. It is evident that

$$m = \rho v - \rho_0 v = v(\rho - \rho_0) = m_0 \left(1 - \frac{\rho_0}{\rho}\right) \tag{3.20}$$

where ρ and ρ_0 are the densities of the dispersed phase and disperse medium respectively, and m_0 is the particle mass in vacuum. Disregarding the initial acceleration, and assuming that the particle moves at a uniform rate, we obtain according to (3.13)

$$mg = uB, \tag{3.21}$$

where $u = \mathrm{d}h/\mathrm{d}t$ is the velocity of sedimentation.

For spherical particles [$m = \frac{4}{3}\pi r^3(\rho - \rho_0)$ and $B = 6\pi\eta r$] we have

$$u = \frac{2}{9}\frac{r^2}{\eta}(\rho - \rho_0)g. \tag{3.22}$$

It follows from (3.22) that the sedimentation velocity is directly proportional to the square of particle radius, and inversely proportional to the viscosity of the disperse medium; u also depends on $(\rho - \rho_0)$ in such a way that for $\rho > \rho_0$ sedimentation takes place, whereas at $\rho < \rho_0$ (*e.g.* emulsion of oil in water) the particles rise to the surface (inverse sedimentation or creaming).

When the system is acted upon by an artificial centrifugal field whose acceleration $\omega^2 x$ (ω is the angular velocity, and x the distance from the axis of rotation) is so much greater than g that the latter can be disregarded, sedimentation takes place in the artificial centrifugal field. The velocity of such sedimentation is determined by the equation

$$m\omega^2 x = u(x)B. \tag{3.23}$$

This formula differs from (3.21) in that $u(x)$ is a function of x, whereas in (3.21) u is independent of the height. The magnitude

$$S = \frac{u(x)}{\omega^2 x} = \frac{m}{B} \tag{3.24}$$

is called the *sedimentation constant* and is determined by the particle mass and (through B) by its shape.

The above formulae show that sedimentation studies lead to the determination of m/B, and in the case of spherical particles, to the determination of their diameter.

References p. 74–75

It is not difficult to calculate the minimum radius r_{min} of particles which settle under the influence of gravity. If we assume, for example, that the minimum perceptible sedimentation velocity is 1 cm/h, or 1/3600 cm/sec, then for a hydrophilic colloid of density $(\rho-\rho_0)\approx 1$, sedimenting in a medium of viscosity $\eta = 10^{-2}$ poise under the influence of gravity ($g = 981 \approx 10^3$ cm/sec^2), we find

$$r_{min} \approx 10^{-4} \text{ cm} = 1\ \mu.$$

To bring about the sedimentation of finer particles it is necessary to use higher centrifugal fields. Ultracentrifuges are generally used for this purpose, in which very high accelerations may be produced (see below).

Conventional sedimentation is made use of in the so-called sedimentation analysis, in which the distribution of particle sizes in a polydisperse system, *e.g.* in powder suspensions, is determined. To this end the powder is dispersed in water or another liquid, and sedimentation of the powder is investigated as a function of time. Several methods of observation have been developed.

1. Wiegner's method (1918). This method is based on the variation of the mean specific gravity of a suspension when, as a result of sedimentation, the powder in suspension has settled. Wiegner's apparatus is shown in Fig. 16. The suspension is placed in the wide tube A, and the pure solvent in the narrow tube B; both limbs are connected by opening the stopcock. The level in B is higher by h than that in A, since the mean specific gravity of the liquid in arm A is greater than that in arm B. As the powder in suspension settles, the specific gravity of the suspension approaches that of the pure solvent, and h tends towards zero*. The dependence of h on the time of sedimentation enables the sedimentation velocity and hence the size of the settling particles to be determined.

2. Methods in which the quantity of substance which has settled during sedimentation is determined by weighing are capable of a high order of accuracy. Sven Odén (1919)[13] developed a very accurate, although complex, apparatus for such measurements. A very convenient and simplified hydrostatic balance was proposed for the same purpose by Figurovskii[14]. A light glass cup is suspended from the end of an elastic quartz or glass thread arranged horizontally, the deflection of which permits the

* When a correction for capillary is taken into account.

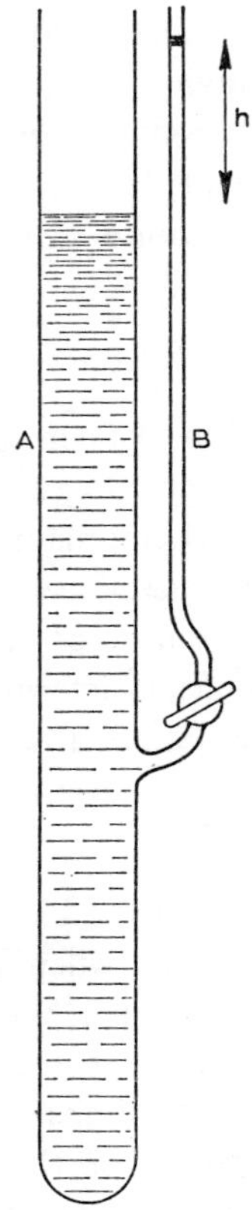

Fig. 16. Wiegner's apparatus for sedimentation analysis.

weight of the substance settling in the cup to be estimated. The cup is immersed in a cylinder filled with the suspension. Deflection of the elastic filament is determined periodically by means of a horizontal microscope, and the quantity of substance which precipitates on the cup is calculated as a function of time.

6. Sedimentation and Diffusion. Sedimentation Equilibrium

Let us now analyse the case in which both sedimentation and diffusion must be taken into consideration if the behaviour of the disperse system is to be correctly described.

If u is the sedimentation velocity and c the concentration, then $uc = i_s$ is the sedimentation flux which is expressed as follows for the usual sedimentation (it is assumed that the x axis of the coordinate system is vertical):

$$i_s = -\frac{mg}{B}c, \tag{3.25}$$

and for centrifugal sedimentation:

$$i_s(x) = \frac{m\omega^2 x}{B}c \tag{3.26}$$

Comparison of the sedimentation flux i_s with the diffusion flux i_D provides a convenient method of determining whether sedimentation or diffusion is the predominating process in the given case. When $|i_s/i_D| \gg 1$ sedimentation predominates, whereas in the case of $|i_s/i_D| \approx 1$ both processes must be taken simultaneously into consideration*. The ratio i_s/i_D can be easily obtained from (3.2) and (3.25)

$$\frac{i_s}{i_D} = \frac{mg}{kT} \cdot \frac{c}{\mathrm{d}c/\mathrm{d}x} = -\frac{v(\rho - \rho_0)g}{kT} \cdot \frac{c}{\mathrm{d}c/\mathrm{d}x} \tag{3.27}$$

for the case of common sedimentation, and according to (3.26)

$$\frac{i_s}{i_D} = -\frac{m\omega^2 x}{kT} \cdot \frac{c}{\mathrm{d}c/\mathrm{d}x} = \frac{v(\rho - \rho_0)\omega^2 x}{kT} \frac{c}{\mathrm{d}c/\mathrm{d}x} \tag{3.28}$$

for sedimentation in an artificial centrifugal field.

These equations indicate that for a given difference of densities i_s/i_D increases with increasing particle size (through the particle volume v). Sedimentation plays therefore a greater role in case of systems with a coarser dispersion. On the other hand, and this is an essential feature, when we start with a system possessing a uniform distribution, sedimentation always predominates at the outset since $\mathrm{d}c/\mathrm{d}x = 0$. After some time, as the sedimentation process progresses, the uniform distribution of the substance is disturbed, $\mathrm{d}c/\mathrm{d}x$ becomes finite and increases with time, and i_D increases until it becomes equal to i_s. At this point no further mass transfer takes place and equilibrium is established. The equilibrium state* is thus characterized by $i_s/i_D = -1$ and it is hence possible to calculate the equilibrium distribution of concentration in the terrestrial gravitational or in an artificial centrifugal field. In the former case, having substituted dh

* When equilibrium has been achieved, there is no transfer of material *i.e.*, the net flow $i_s + i_D = 0$.

for dx, we obtain (since the gradient is produced along the height)

$$-\frac{mg}{kT}\frac{c}{\mathrm{d}c/\mathrm{d}h}=1\,.$$

After separating the variables and integrating between 0 and h and correspondingly between c_0 at $h=0$ and c_h at h, we obtain

$$\ln\left(\frac{c_0}{c_h}\right)=\frac{mgh}{kT} \tag{3.29}$$

i.e. the barometric equation.

Similarly, for the artificial centrifugal field

$$\frac{m\omega^2 x}{kT}\frac{c}{\mathrm{d}c/\mathrm{d}x}=1\,.$$

After integration between c_1 and c_2 and, correspondingly, from x_1 to x_2, we obtain

$$\ln\left(\frac{c_1}{c_2}\right)=\frac{m\omega^2}{2kT}(x_1^2-x_2^2)\,. \tag{3.30}$$

The equilibrium distribution of substance in the volume, just as any other equilibrium, does not depend on the way in which this distribution had been reached. The method used here for calculating this distribution (making the diffusion flux equal to the sedimentation flux) is only one of the possible calculation methods. The occurrence of mass transfer in this case is not the indispensable condition, since the equations obtained do not include the magnitude B which characterizes the transfer and the accompanying friction. The equilibrium distribution can be calculated without taking into consideration the mechanism of its establishment. This mechanism is important only with respect to the time required to attain the equilibrium distribution or, more exactly, the time during which a distribution close to the equilibrium one is established (the time required for the establishment of the true equilibrium is always infinitely long).

The thermodynamic derivation is based on Boltzmann's law, according to which at equilibrium the relative probabilities of particles being found in two places, characterized by the ratio of the numbers of particles (particle concentrations) in these places, are given by the ratio of magnitudes $\mathrm{e}^{-u/kT}$, where u is the potential energy of particles in the places in question. Therefore,

$$\frac{c_1}{c_2}=\frac{\mathrm{e}^{-u_1/kT}}{\mathrm{e}^{-u_2/kT}}=\mathrm{e}^{-\Delta u/kT}\,. \tag{3.31}$$

References p. 74–75

The potential energy difference $\Delta u = u_1 - u_2$ for places situated at distances x_1 and x_2 from the axis of rotation is equal to the work of the reversible isothermal transfer of one particle from x_1 to x_2. In the case of common sedimentation $du = mg \cdot dh$ and $\Delta u = mgh$; after substituting Δu in (3.31) we can hence obtain (3.29)*.

Eqn. (3.29) was used by Perrin for the determination of Boltzmann's constant and thus of Avogadro's number. To this end Perrin determined the number of particles at various levels in monodisperse gamboge suspensions with a known value of m. The investigation was carried out in a micro-cell which had been allowed to stand for several weeks after filling with the suspension to make completely sure that the equilibrium had been established. The N_A values obtained in this way were in good agreement with values obtained by other methods. Certain authors question however the validity of these measurements; it is in particular uncertain whether equilibrium is really attained in the system, because of convection and other disturbances.

The ultracentrifuge

In 1912 Dumanskii suggested that use should be made of artificial centrifugal fields for the sedimentation of colloidal particles. This author was also the first to apply this method, using relatively small accelerations. In the following year Svedberg[15] developed an ultracentrifuge for high accelerations, and carried out quantitative measurements of sedimentation and sedimentation equilibria, thus creating a now widely used method of investigating colloidal and high-molecular solutions in the ultracentrifuge.

The modern ultracentrifuge is of a complex design. In the highest-power ultracentrifuges the rotor, which is a few centimeters in diameter and is made from a high-strength material, rotates in a vortex of hydrogen. Hydrogen ensures a rapid withdrawal of heat generated by friction, and reduces heat convection. Several plane quartz cells filled with the solution to be examined are fixed radially in the rotor. The centrifuge casing is fitted with small quartz windows through which the rotating cells can be observed in transmitted light each time the cell traverses the light beam

* The method given above may also be used to prove (3.8): at equilibrium in an arbitrary field, acting on a particle with force f we have $i = -D \frac{dc}{dx} + \frac{f}{B} c = 0$. Since $f = -\frac{du}{dx}$, the solution has the form $c = c_0 e^{-u/BD}$, which, on comparison with Boltzmann's law $c = c_0 e^{-u/kT}$, gives $D = kT/B$.

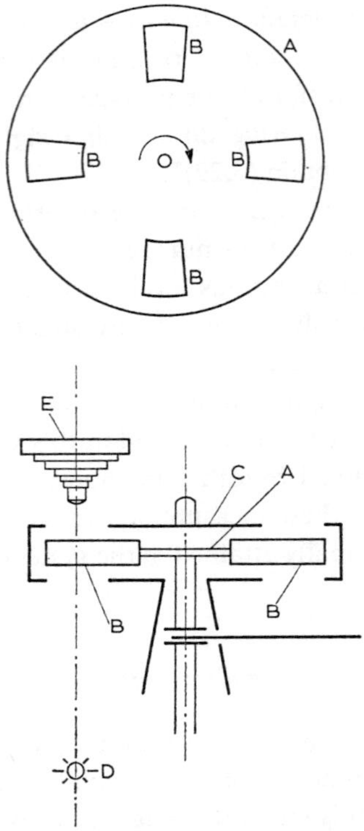

Fig. 17. Diagrammatic layout of the ultracentrifuge: (A) rotor, (B) cells, (C) centrifuge cover, (D) source of light, (E) photographing apparatus.

which is transmitted through the windows and is allowed to fall on to a photometer or a photographic plate. In the latter case the photograph is photometered. The intensity of light which has passed through the solution examined is determined as a function of its concentration in a simultaneous separate experiment. It is thus possible to obtain, from the results of photometry, information regarding the concentration distribution in the cell at the moment of photometering or photographing. If the photometering or photographing is effected at certain intervals, it is also possible to follow the sedimentation process in time. The layout of the ultracentrifuge is given in Fig. 17 which shows an ultracentrifuge not capable of attaining very high speeds, its rotor being rotated by means of a motor*.

* Translator's note: the high-speed ultracentrifuge is driven by a built-in air turbine.

The optical methods available for the determination of the concentration distribution during sedimentation or of the steady distribution at equilibrium are fairly complex. To this end use is generally made of the dependence of the refractive index on concentration. This method is also used in other cases (determination of the diffusion coefficient, and in the Tiselius' method of electrophoresis, in the description of which the method will be treated in greater detail).

Accelerations amounting to hundreds of thousands of gravities have been attained in modern ultracentrifuges at speeds of the order of 10^3 rev/sec; this results in a clearly measurable sedimentation velocity of macromolecules.

Ultracentrifuges are used for the investigation of sedimentation, as well as of the steady-state concentration distribution.

In the latter case lower accelerations are required (speeds of the order of 10^2 rev/sec). When the equilibrium state is attained, the concentration distribution obeys (3.30). Assuming that in this equation $m = m_0 \{1 - (\rho_0/\rho)\} = M/N_A \{1 - (\rho_0/\rho)\}$, we obtain for the molecular weight the relation

$$M = \frac{2RT \ln c_1/c_2}{(1 - \rho_0/\rho)\,\omega^2(x_2^2 - x_1^2)} \qquad (3.32)$$

where c_1 and c_2 are concentrations at distances x_1 and x_2 from the axis of rotation. According to the derivation of (3.30), this equation can be written in the differential form

$$M = \frac{RT}{(1 - \rho_0/\rho)\,\omega^2 x} \frac{\mathrm{d}c_x}{c_x\,\mathrm{d}x}, \qquad (3.33)$$

where c_x is now the concentration at a distance x.

The above equations have been derived assuming that a colloidal solution obeys the equation of state of ideal solutions. Since solutions of high polymers deviate considerably from this state, it is necessary when determining M, either to extrapolate to $c \to 0$ or to introduce corrections.

Solutions of high polymers are usually polydisperse. The resulting sedimentation pattern is consequently the sum of distributions superposed on one another, each corresponding to polymer fractions of different molecular weights. It is readily seen that heavy molecules will predominate away from the axis of rotation. The mean molecular weight determined by formula (3.32) at various zones of x will therefore not be

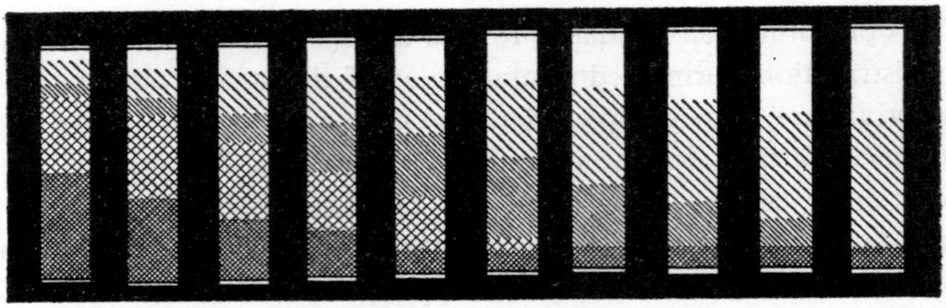

Fig. 18. Successive photographs of cells with haemocyanin solution (containing four components having different sedimentation constants), rotating in the ultracentrifuge.

constant, but will increase towards small values of x. This variation of M with x is a proof of polydispersion. In such cases c_x and dc_x/dx are simultaneously determined for consecutive values of x, and data regarding the distribution of macromolecules with respect to their molecular weights are obtained from these results.

The method of investigating the sedimentation velocity is highly interesting. Given sufficient accelerations, sedimentation prevails over diffusion, and a sharp boundary forms between the solution and solvent zones; the velocity of displacement of this boundary (towards the axis of rotation for $\rho > \rho_0$) yields directly the velocity of sedimentation. When the solution contains several components with different sedimentation constants, then after a certain time, a corresponding number of boundaries is established in the ultracentrifuge cell. It thus becomes possible to determine the number of components and, from the velocity of displacement of each boundary, to identify each component by the corresponding sedimentation constant. This procedure is often used in the case of biological objects* (Fig. 18). To eliminate the effect of shape (the friction constant B) from the expression for the sedimentation constant (3.24) and to obtain M, the diffusion constants $D = kT/B$ of the corresponding components are also determined. Assuming that B is the same for diffusion and sedimentation (this is not absolutely true since the orientating factors, and consequently the predominant orientation of the particle motion may not be the same in both cases), B may be eliminated. Such determinations are certainly feasible when the components may be isolated in pure form for their diffusion to be investigated. However, if it is possible to separate the

* In solutions of proteins the sedimentation constants are of the order of 10^{-13}–10^{-12} sec. $S = 10^{-13}$ sec is called one Svedberg.

components, it might be better to determine their molecular weights directly by the sedimentation equilibrium method.

The possibilities referred to above have contributed to the wide use of the ultracentrifuge method in spite of its technical difficulties and high cost. A detailed though rather out-of-date exposition of this method can be found in "The Ultracentrifuge" by Svedberg and Pedersen[15]. The present author believes that the fundamental possibilities of this method have not as yet been fully exhausted. For instance, if another homogeneous field acting on the particles is applied to a multicomponent solution subjected to a centrifugal acceleration, it is possible to separate components at the sedimentation equilibrium, thus determining at the same time the number of components and their molecular weights. The author demonstrated theoretically (1956)[16] such a possibility for the reversed sedimentation when a homogeneous electric field was used as the second field.

7. Viscosity

Colloidal solutions prepared from high-molecular substances possess a much greater viscosity than that of the pure solvent. Furthermore, in a number of cases their viscosity, unlike that of low-molecular (Newtonian) liquids depends on the velocity of flow, and varies markedly with time on passing from one flow velocity to another. Colloidal solutions are also endowed with other characteristic features with regard to their mechanical properties. In these systems there exists the possibility of a gradual transition (associated with a change of concentration, temperature and other parameters) from Newtonian liquids, through non-Newtonian liquids, to the properties of plasticity, elasticity and hardness, characteristic of solids. These very clearly expressed properties of colloids are extremely important in practice and are also highly interesting from the theoretical point of view. Unfortunately, the theoretical interpretation of these highly complex phenomena is at present still in the stage of searching for a quantitative theory. For this reason, in spite of the importance of this field, it cannot be discussed at the level of this book; we shall therefore confine ourselves to a limited review of viscosity, necessary for the understanding of other points brought out in this and other chapters.

Viscosity, or internal friction during the motion of the liquid, is due to the fact that various layers of moving liquids flow with different velocities and therefore glide with respect to each other. As a result of the random molecular-kinetic movement and of molecular attractive forces, the faster flowing layers tend to carry with them the layers which move more slowly, and *vice versa*. In this way friction takes place between adjacent layers, which gives rise to irreversible conversion of a part of the energy of motion to heat, *i.e.* to dissipation of this energy.

The laws of viscous flow, *i.e.* hydrodynamic equations of motion in which friction is taken into account (Navier–Stokes equations) are highly complicated, and will not be discussed. To explain some problems related to viscous flow, we shall make use of Newton's law, which enables some particularly simple cases to be described.

This law is expressed as follows:

$$f = \eta S \frac{\mathrm{d}u}{\mathrm{d}x} \tag{3.34}$$

where f is the frictional force, S the surface area of layers undergoing friction, in the flow direction y, and $\mathrm{d}u/\mathrm{d}x$ the velocity gradient in the direction at right angles to y. The coefficient η, a constant characteristic of every liquid, is called viscosity and has the dimensions of $[\mathrm{g \cdot cm^{-1} \cdot sec^{-1}}]$ according to (3.34). The c.g.s. unit of viscosity is called the poise, in honour of Poiseuille. For instance, the viscosity of water at room temperature is 10^{-2} poise, or 1 centipoise.

Making use of Newton's law (3.34) let us calculate the rate at which a liquid flows through a tube, *i.e.* let us derive Poiseuille's equation[17], which we shall repeatedly apply and which is the basis of the viscosity determinations by one of the most widely used instruments, the Ostwald viscometer[18].

Let us first determine the velocity u_x at any point in the liquid flowing uniformly and fairly slowly along a tube under the influence of a pressure difference P between the ends of the tube. Under these conditions the flow is laminar, *i.e.* in every liquid layer whose distance from the tube wall is the same, the velocity of flow is the same. Moreover, if the cross-section area of the tube is constant, the pressure is distributed uniformly, satisfying the condition:

$$\frac{\mathrm{d}P}{\mathrm{d}y} = \frac{P}{l},$$

where l is the length of the tube.

Near the walls ($x=r$) the velocity of the flow is zero. Consider an

References p. 74–75

imaginary cylinder of length dy and radius x $(x < r)$ coaxial with the outer one (Fig. 19). According to (3.34) this cylinder experiences a frictional force df given by

$$\mathrm{d}f = \eta\,\mathrm{d}S\frac{\mathrm{d}u}{\mathrm{d}x} = \eta\,2\pi x\,\mathrm{d}y\frac{\mathrm{d}u}{\mathrm{d}x},$$

where $2\pi x$ dy is the surface area of this cylinder.

During uniform flow the frictional force is numerically equal to the force $-\pi x^2 \mathrm{d}P$, due to the pressure which determines the movement, so that

$$\frac{1}{x}\frac{\mathrm{d}u}{\mathrm{d}x} = -\frac{1}{2\eta}\frac{\mathrm{d}P}{\mathrm{d}y} = -K.$$

Since K is constant,

$$\mathrm{d}u = -Kx\,\mathrm{d}x \quad \text{and} \quad u = -\frac{K}{2}x^2 + \text{const.}$$

The boundary condition at $x = r$ is $u = 0$, so that

$$u = -\frac{K}{2}x^2 + \frac{K}{2}r^2,$$

or, after substituting for K,

$$u = \frac{1}{4\eta}\frac{\mathrm{d}P}{\mathrm{d}y}(r^2 - x^2).$$

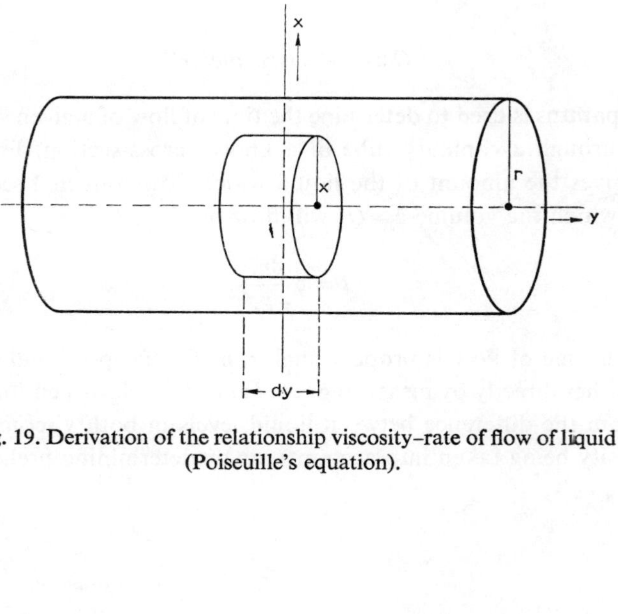

Fig. 19. Derivation of the relationship viscosity–rate of flow of liquid (Poiseuille's equation).

Furthermore, by substituting P/l for $\mathrm{d}P/\mathrm{d}y$, we obtain

$$u = \frac{1}{4\eta}\frac{P}{l}(r^2 - x^2). \tag{3.35}$$

This equation shows that velocity of flow at a distance x from the centre depends parabolically on x, being zero at the walls and maximal along the axis ($x=0$):

$$u_{\text{max}} = \frac{1}{4\eta}\frac{P}{l}r^2 .$$

From eqn. (3.35) it is also possible to calculate the quantity Q of the liquid, which flows out of the tube in 1 sec. For this purpose it is first necessary to calculate $\mathrm{d}Q$, the quantity of liquid transported by a hollow cylindrical layer of radius x and a wall thickness $\mathrm{d}x$. According to the above, the velocity of flow in this layer is u, and

$$\mathrm{d}Q = u2\pi x\,\mathrm{d}x = \frac{P\pi}{2\eta l}(r^2 x\,\mathrm{d}x - x^3\,\mathrm{d}x).$$

By integrating $\mathrm{d}Q$ for all layers from $x=0$ to $x=r$, the quantity of liquid which flows out during a unit of time:

$$Q = \frac{\pi P r^4}{8\eta l}. \tag{3.36}$$

This is the well-known Poiseuille equation for viscous flow through a tube.

Ostwald's viscometer[18]

This apparatus is used to determine the time of flow of a given volume of a liquid through a capillary tube of a known cross-section. From (3.36), which gives the amount of the liquid which flows out in 1 sec, the time during which the volume $v = Qt$ will flow out is:

$$t = v\frac{8\eta l}{\pi P r^4}.$$

Thus the time of flow is proportional to η. The proportionality factor is found either directly by measuring r, v, l and P (the last-mentioned magnitude from the difference between liquid levels in both viscometer limbs, the density being taken into account), or by determining preliminarily in

References p. 74–75

the same viscometer the time of flow of a liquid with a known viscosity. Since viscosity is highly dependent on temperature, the measurements must be carried out in a thermostat.

The falling ball method

In this case use is made of Stokes' formula according to which the force of friction f of a sphere during its movement in a viscous liquid is

$$f = 6\pi\eta r,$$

where r is the radius of the sphere.

If the ball falls slowly and if the vessel filled with the liquid in which the ball falls is sufficiently wide (conditions at which the above equation is valid), we can write

$$u = \frac{mg}{6\pi\eta r}$$

where m is the effective mass of the ball in the corresponding liquid, g is the acceleration due to gravity, and u is the velocity at which the ball is falling. Thus the velocity of fall is inversely proportional to the viscosity. This proportionality is also maintained in the case when the ball moves not in a wide vessel, but in a slanting tube (Hoppler's viscometer), although in this case the frictional force is described by an expression which is more complex than Stokes' equation. When the viscosity is determined by comparing the time of fall of the ball in the liquid tested with that in a liquid of known viscosity, the expression for the frictional force becomes superfluous.

Couette's viscometer[19]

This viscometer is particularly convenient for the investigation of the variation of viscosity with time, which is very frequently observed in colloidal systems, *e.g.* as a result of coagulation. This viscometer consists of a cylinder suspended by a thin elastic filament to which a mirror for determining the torsion angle is attached. The cylinder is placed coaxially in a cylindrical vessel holding the liquid under examination. The outer cylinder rotates at a constant speed. This motion is transferred to the investigated liquid which in turn sets in motion the inner cylinder, turning it round its axis until the torsional force is balanced by the force of friction. It was demonstrated that the torsion angle is proportional to the viscosity, and that therefore, the viscosity of one liquid can be determined, if the viscosity of another liquid is known, by comparing the torsion angles in the two liquids for the same rate of rotation.

8. Viscosity of Suspensions of Spherocolloids

It has already been pointed out that the viscosity of colloidal solutions always exceeds that of the pure solvent. This increase of viscosity is expressed less sharply in dilute solutions, when the interaction of particles and their possible collisions do not play a substantial role. In 1906 Einstein[10b] analysed in detail the extreme case of a solvent containing solid spherical particles of identical size. In such a case the absence of interaction between the particles implies not only the absence of static force acting between the particles, but also the absence of dynamic interactions caused by the movement of particles. In other words, Einstein's model was a suspension in which the distances between the particles were so considerable that the movement of each particle could be regarded as a movement of one particle in an infinitely large volume of the liquid.

For this case Einstein derived the following equation:

$$\eta = \eta_0(1 + 2.5\varphi)$$

where η_0 is the viscosity of the solvent, and φ the ratio of the volume of the dispersed substance to that of the solution. This equation displays a particularly interesting feature, *i.e.* that the viscosity is independent of the degree of dispersion.

Einstein's calculations are highly complicated, and will not be given here. An approximate derivation is however presented below to provide a certain idea of the physical nature of the phenomenon, the inaccuracy of which is due to certain simplifications.

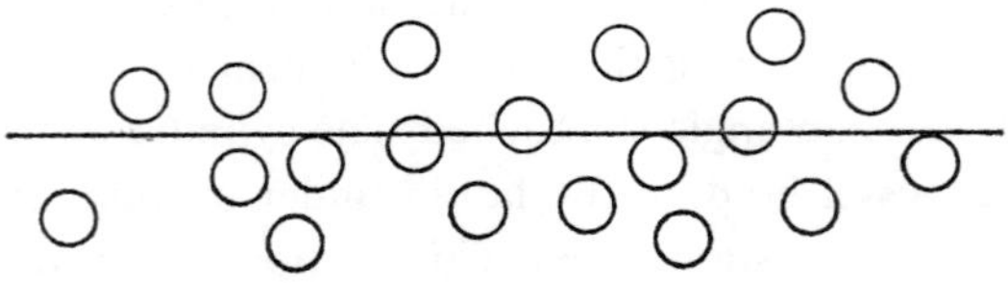

Fig. 20. Cross-section through a stream of a viscous liquid.

References p. 74–75

Let us consider a cross-section of a stream of a viscous liquid which contains a known quantity of rigid spherical particles of identical dimensions. During the flow of the liquid its two adjacent layers glide with respect to each other (Fig. 20). It is clear that the friction will be caused both by drag f_1 between the liquid regions free from colloidal particles, and by drag f_2 which arises when the liquid flows around the particles. The total frictional force is therefore

$$f = f_1 + f_2 \, .$$

If the analysed area is 1 cm^2, we can write for f_1, in accordance with Newton's law,

$$f_1 = \eta_0 (1 - S_1) \frac{\mathrm{d}u}{\mathrm{d}x} \, ,$$

where η_0 is the viscosity of the solvent, and S_1 the total area of the sections through the individual particles. The part $\mathrm{d}S_1$ of this occupied area, which corresponds to particles whose centres are situated at a distance between x and $x + \mathrm{d}x$ from the slip plane, is given by

$$\mathrm{d}S_1 = \pi (R^2 - x^2) c \, \mathrm{d}x \, ,$$

where R is the particle radius, and c the number of particles in a unit volume. Correspondingly,

$$S_1 = 2 \int_0^R \pi (R^2 - x^2) c \, \mathrm{d}x = \tfrac{4}{3} \pi R^3 c$$

will be the sum of all cross-sections of particles intersected by the friction plane. Hence

$$f_1 = \eta_0 (1 - \tfrac{4}{3} \pi R^3 c) \frac{\mathrm{d}u}{\mathrm{d}x} \, .$$

Similarly,

$$f_2 = \eta_0 S_2 \left(\frac{\mathrm{d}u}{\mathrm{d}x} \right)_0 \, ,$$

where S_2 is the streamline surface of all particles intersected by the indicated plane:

$$S_2 = 4 \pi R^2 \cdot 2Rc = 8 \pi R^3 c \, ,$$

and $(\mathrm{d}u/\mathrm{d}x)_0$ is the mean velocity gradient near the surface of these particles.

If the particles entrained by the liquid move without rotation, the gradient $(\mathrm{d}u/\mathrm{d}x)_0$ would be of the same order of magnitude as $\mathrm{d}u/\mathrm{d}x$. In reality however $(\mathrm{d}u/\mathrm{d}x)_0 < \mathrm{d}u/\mathrm{d}x$, since particles rotate owing to the different velocities of layers which flow around them. In this case the velocity gradient and consequently also the friction decreases, for the same reason as in the case of ball bearings. This effect cannot be allowed for quantitatively in the present elementary derivation, and it is therefore assumed that

$$\left(\frac{\mathrm{d}u}{\mathrm{d}x}\right)_0 = \frac{\mathrm{d}u}{\mathrm{d}x}.$$

Rotation is thus neglected, and the values obtained for the viscosity of suspensions are too high. Then for f_2 we have

$$f_2 = \eta_0 8\pi R^3 c \frac{\mathrm{d}u}{\mathrm{d}x},$$

so that

$$f = f_1 + f_2 = \eta_0(1 + \tfrac{20}{3}\pi R^3 c)\frac{\mathrm{d}u}{\mathrm{d}x}.$$

Taking into account that $c = N/V$ (where V is the volume of solution, and N the number of all particles in this volume) and that $\frac{4}{3}\pi R^3 N = v$ is the volume of the dispersed substance, we obtain

$$f = \eta_0(1 + 5\varphi)\frac{\mathrm{d}u}{\mathrm{d}x},$$

where $\varphi = v/V$.

The force f can be represented (in this case $S = 1$) as follows, according to Newton's law

$$f = \eta \frac{\mathrm{d}u}{\mathrm{d}x},$$

where η is the viscosity of the suspension. We have then

$$\eta = \eta_0(1 + 5\varphi)$$

i.e. we obtain an expression similar to Einstein's equation, but the numerical coefficient in front of φ is in this expression not 2.5 but 5. As already indicated, this is due to the simplification made in the derivation of the equation, when the rotation of particles was disregarded.

Einsteins' equation can be written as

$$\frac{\eta - \eta_0}{\eta_0} = 2.5\varphi.$$

References p. 74–75

The relative increase of the viscosity of suspension $(\eta - \eta_0)/\eta_0$ is often called, not very accurately, the *specific viscosity*.

Einstein's equation was repeatedly checked experimentally. Such experiments are very difficult, since conditions at which this theory is valid are fulfilled only in very rare cases. A good agreement with the theory was obtained by Eirich in 1936–37[20], during an investigation of the viscosity of suspension of glass spheres and of spherical spores in highly diluted solutions.

The extreme simplification of Einstein's assumptions, which is the cause of the disagreement between experimental and theoretical results, has prompted further development of the theory for more complex cases.

The simplest correction was put forward by Dumanskii in 1913[21]. If the particle combines a certain quantity of the dispersion medium in the form of a solvated layer strongly retained by the surface of the particle, then this solvation must also be allowed for in the expression for φ when the volume v of the dispersed phase is calculated. If the thickness of the solvated layer is constant, then φ, and consequently η, is found to depend on the degree of dispersion; if only solvation is responsible for this deviation from Einstein's equation, we can then use this equation to calculate the amount of the solvation-bonded dispersion medium.

Einstein's equation is not valid for more concentrated solutions, primarily because of the arising hydrodynamic interactions. For this case the expression

$$\frac{\eta - \eta_0}{\eta_0} = a_1\varphi + a_2\varphi^2 + a_3\varphi^3 + \ldots$$

may be used, where a_1, a_2, ... are calculated on the basis of a more complex theory. It is clear that a_1 must be 2.5 if we are to obtain Einstein's equation at infinite dilution ($\varphi \to 0$). In 1936 Guth and Simha[22] obtained theoretically $a_1 = 2.5$ and $a_2 = 14.1$. The experimental data are not however in agreement with theoretical predictions. Thus, Eirich found $a_1 = 2.5$ and $a_2 = 7.8$, and Eilers (1941–43)[23] found for bitumen suspensions in paraffin oil $a_1 = 2.5$, $a_2 = 4.94$ and $a_3 = 8.78$, *i.e.* values lower than those predicted by the theory.

A highly important and complex problem is that of the viscosity of a suspension of anisodiametric particles. In this case there occurs a competition between the orientating effect of the flow and the disorientating effect of Brownian movement. In the case of small velocities of flow and small particles Brownian movement plays an important role, which can even result in complete disturbance of hydrodynamic orientation. In the case of high velocities of flow and large particles, viscous friction takes place under conditions of optimum hydrodynamic orientation, and Brownian movement can be disregarded. Attempts at solving these complicated problems [*e.g.* Jeffery (1923)[24]–without allowing for Brownian movement; Eisenschitz (1931–33)[25]; Kuhn (1945)[26] and others –allowing for Brownian movement] have so far failed to yield reliable results. It is interesting to note that Brownian movement increases the viscosity by disturbing the preferred (lowest friction) orientation of particles in the stream. This is therefore a case in which the viscosity is a function of the velocity of flow (the role of Brownian disorientation decreases at high velocities), *i.e.* the case of the so-called non-Newtonian liquids.

The dependence of viscosity on the velocity of flow increases much more when a structure appears in a disperse system. For reasons explained in the introduction to this section we shall not deal with these important problems.

In spite of the absence of a reliable theory, viscosity measurements are widely used in solutions of macromolecular substances, as a method of determining their molecular weights. The following semi-empirical equation is used in these cases:

$$\lim \frac{\eta - \eta_0}{\eta_0 c} = KM^\alpha,$$

where K and α are determined experimentally by using the absolute methods of measuring M. It is frequently found that these constants remain unchanged for substances of similar chemical structure.

REFERENCES

1 FICK, *Pogg. Ann.*, 94 (1855) 59.
2 FRENKEL, *Z. Physik*, 35 (1926) 652.
3 FUOSS AND MEAD, *J. Phys. Chem.*, 47 (1943) 59.
4 SCHULZ in: STUART (Herausgeber), *Die Physik der Hochpolymeren*, Springer, Berlin, 1953, Vol. 2.
5 CH'IEN, *Opredelenie molekulyarnykh vesov polimerov* (*Determination of the molecular weights of polymers*), Izdatel'stvo inostrannoi literatury, Moscow, 1962.
6 DONNAN, *Z. Elektrochem.*, 17 (1911) 572.
7 NERNST, *Z. Physik. Chem.* (*Leipzig*), 2 (1888) 612.
8 BROWN, *Phil. Mag.*, 4 (1828) 161.
9 ECHSNER, *Ann. Physik*, 2 (1900) 843.

10a EINSTEIN AND SMOLUCHOWSKI, *Brownian movement, collection of papers*, with comments by KRUTIKOV AND DAVYDOV, ONTI, Moscow, 1936.
b EINSTEIN, *Ann. Physik*, 17 (1905) 549; 19 (1906) 371; 19 (1906) 289; *Z. Elektrochem.*, 13 (1907) 41; 14 (1908) 235.
c SMOLUCHOWSKI, *Bull. Intern. Acad. sci. Cracovie*, (1906) 202; *Ann. Physik*, 21 (1906) 756; *Physik. Z.*, 13 (1912) 1069; 14 (1913) 261; *Bull. intern. Acad. sci. Cracovie, A* (1913) 418; *Sitz.-ber. Akad. Wiss. Wien*, 123 (1914) 2381; 124 (1915) 263, 339; *Ann. Physik.* 48 (1915) 1103; *Physik. Z.*, 16 (1915) 318; 17 (1916) 557.
11 PERRIN, *Ann. Chim. Phys.*, 18 (1909) 5; *Physik. Z.*, 11 (1911) 461.
12 LANGEVIN, *Compt. Rend.*, 146 (1908) 530.
13 ODÉN, *Kolloid-Z.*, 26 (1919) 100.
14 FIGUROVSKIĬ, *Sedimentometricheskiĭ analiz*. Akad. Nauk SSSR, Moscow, 1948.
15 SVEDBERG, PEDERSEN *et al. The ultracentrifuge*. Oxford University Press, London 1940.
16 SHELUDKO, *J. Colloid Sci.*, 11 (1956) 167.
17 POISEUILLE, *Compt. rend.*, 11 (1840) 961, 1041; 15 (1842) 1167.
18 OSTWALD, *Lehrbuch der allgemeinen Chemie*, Leipzig, 1891, p. 580.
19 COUETTE, *Ann. phys. et chim.*, 21 (1890) 433.
20 EIRICH, BUNZL AND MARGARETHA, *Kolloid-Z.*, 74 (1936) 275; EIRICH AND GOLDSCHMIDT, *Kolloid-Z.*, 81 (1937) 7.
21 DUMANSKII, *Kolloid-Z.*, 12 (1913) 6.
22 GUTH AND SIMHA, *Kolloid-Z.*, 74 (1936) 266.
23 EILERS, *Kolloid-Z.*, 97 (1941) 313; 102 (1943) 154.
24 JEFFERY, *Proc. Roy. Soc. (London)*, A 102 (1923) 163.
25 EISENSCHITZ, *Z. Physik. Chem.*, A 158 (1931) 78; A 163 (1933) 133.
26 KUHN AND KUHN, *Helv. Chim. Acta*, 38 (1945) 97.

CHAPTER IV

Physical Chemistry of Surfaces

Heterogeneous systems differ from homogeneous ones in that phase interfaces are present. The properties of substances and the distribution of components in a thin layer adjacent to such a phase boundary differ from their bulk values. These peculiarities are reflected in the general properties of the system to an extent which increases with the interfacial area *i.e.* with the degree of dispersion; they are therefore more important for colloidal systems than for coarse suspensions. The increase of the area of the phase interface with increasing degree of dispersion can be illustrated as follows. Let us assume that the phase to be dispersed is a sphere of radius R; the corresponding volume will be $V_0 = \frac{4}{3}\pi R^3$ and the surface area $S_0 = 4\pi R^2$. After this sphere has been dispersed to form N small spheres of radius r, the total surface area will be $S = 4\pi r^2 N$, for the same volume $V_0 = \frac{4}{3}\pi r^3 N$. Thus $S/S_0 = R/r$. If the radius of the initial sphere R is 1 cm, and if after the dispersion particles of colloidal dimensions with $r = 10^{-6}$ cm are obtained, then the area of the interface will increase by a factor of 10^6.

Modern colloid chemistry is based on the physical chemistry of surfaces to the same extent as on the idea that colloidal systems are systems having a high degree of dispersion; thus the combination of these two notions produces the single basis of the theory of this branch of science. Freundlich[1], in his book, was the first to provide a systematic and very profound exposition of colloid science from the point of view of the physical chemistry of surfaces. Numerous authors have dealt specifically with surface chemistry. With regard to its application in colloid chemistry, the well known book by Adam[2], and another by Adamson[3] are quite successful. In this chapter we shall attempt to expound briefly the essentials of surface chemistry of interest in the science of colloids.

References p. 139–140

1. Thermodynamics of Heterogeneous Systems

As may be remembered from general chemistry and physics, heterogeneous disperse systems carry a high excess of surface free energy, owing to their highly developed interface area, and are therefore thermodynamically unstable. Although in further discussion of this problem it will be shown that this is not absolutely true, in a large number of heterogeneous disperse systems these ideas do not give rise to objections. It may be asked to what extent one is entitled to apply thermodynamic ideas of phase equilibrium to these non-equilibrium systems. If as a result of kinetic factors the heterogeneous disperse system has a certain stability which enables it to exist in the form of the disperse system for a long period of time, which is sufficient for the components to be distributed by molecular transfer (*e.g.* by diffusion) in the interior and in the phase boundaries according to the requirements of the conditions of equilibrium, then it is obvious that the state obtained can be successfully analysed on the strength of corresponding thermodynamic concepts. It will be shown in the analysis of the problem of the stability of lyophobic colloidal solutions that such stability really occurs; this explains the wide occurrence of such systems both in nature and technology. The formation of such relative equilibrium states in colloidal systems results in the fact that, if the colloidal particles consist of a liquid or gas dispersed in a liquid or of a liquid dispersed in a gas, the thermodynamic requirement with respect to particle shape is also satisfied, *i.e.* prevalence of the form with the least surface area, in the simplest case the sphere. By confining ourselves to these arguments, which speak for the applicability of the thermodynamics of equilibrium states to heterogeneous disperse systems, let us now analyse the thermodynamics of heterogeneous systems.

The state of the substance near an interface differs from that in the bulk of a phase. In the immediate vicinity of the interface the molecules interact with adjacent identical molecules and simultaneously with molecules which form the other phase. This is of fundamental importance, since the magnitude of molecular forces depends both on the type of molecules and on the intermolecular distances, *i.e.* on density. Therefore, in the general case the force of attraction between the molecules of one phase differs from that between the molecules of the other phase.

The extra free energy in a surface stems from the unbalance of intermolecular forces in the surface. Molecules in a liquid/gas interface for example are attracted toward the liquid but hardly toward the gas phase in view of its low density. In the interior of the liquid the forces acting on a molecule are fully balanced. An analogous situation exists at the liquid/liquid and liquid/solid interfaces. The total (free) energy of the system therefore equals the bulk (free) energy plus the surface or interfacial (free) energy, the latter term being obviously proportional to the surface or interfacial area.

As well as having additional energy, surfaces are characterized by the fact that in multicomponent systems the concentrations of dissolved materials near the interface often differ from their bulk values (*adsorption*). If adsorption takes place spontaneously the surface free energy must decrease.

In the quantitative description of these phenomena one meets a substantial uncertainty. We have said that near the interface the state of molecules differs from that within the phases but failed to give more precise information regarding the distances from the surface involved. Such more precise information is difficult to obtain since the state of molecules changes gradually as the distance from the surface is decreased. However, owing to the fact that the intermolecular forces vary inversely with a high power (usually the seventh) of the distance and that over a distance of a few molecular diameters they become infinitesimally small, it can be asserted that the characteristic state of the bulk of the phase is generally established at a very short distance from the interface. Consequently, in phases in which the greater part of the substance is sufficiently removed from the surface one can distinguish between a part characterized by the "bulk" properties and another part characterized by "surface" properties. Such delimitation is of course purely schematic, especially in the case of a surface layer in which properties change sharply with the distance from the interface. Moreover, the very concept of an interface as a mathematical plane is scarcely applicable to the surface layer, owing to its discrete molecular structure which, strictly speaking, renders the position of the boundary plane indefinite.

In his fundamental investigation of equilibria in heterogeneous systems (1875–1878) Gibbs[4] succeeded in overcoming all these difficulties and developed a consistent thermodynamic theory which took into account magnitudes characterizing not only the "bulk" properties, but also properties relating to the surface.

To avoid the difficulties associated with the discrimination of properties

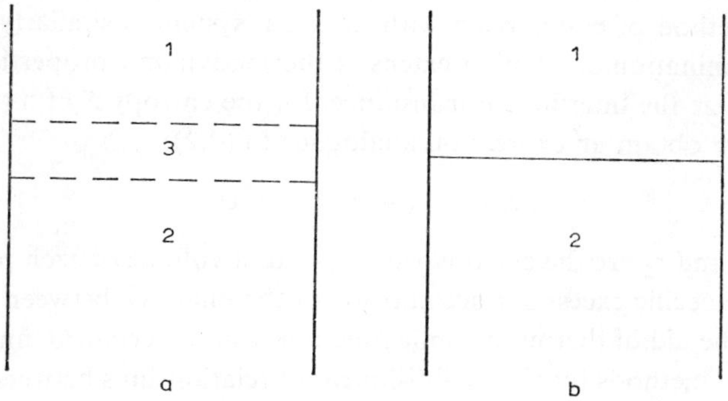

Fig. 21. (a) real system, (b) idealized system.

which are characteristic of interfaces and of those characteristic of the bulk phases, Gibbs suggested the following method. Let us imagine, together with the real system in which properties change more or less gradually on transition from one phase to another (Fig. 21a), an ideal system, in which the properties are identical throughout the entire volume up to the mathematical plane which divides the phases and where they change abruptly into the properties of another phase. It is obvious that the energy of the ideal system (Fig. 21b) which consist of two phases 1 and 2, divided by this plane and having volumes V_1 and V_2 respectively, is equal to the sum of energies of both phases U_1 and U_2,. The energies U_1 and U_2 are in turn determined by the products $u_1 V_1$ and $u_2 V_2$, where u_1 and u_2 are energies per unit volume in phases 1 and 2. The total energy of the ideal system is therefore

$$U_{\text{id}} = u_1 V_1 + u_2 V_2 \,. \tag{4.1}$$

The energy of a real system differs from that of the ideal system, since in the transition region 3 we encounter different properties; in this region the energy per unit volume is different from u_1 and u_2, and the total energy of this system thus exceeds the sum $u_1 V_1 + u_2 V_2$ by a supplementary magnitude U^* which, as has already been mentioned, is proportional to the area of the interface. Consequently $U^* = u^* O$ where u^* is the specific surface energy and O the total interfacial area. It is obvious that u^* is determined by the properties of the substances which form phases 1 and 2. Finally the total energy of the real two-phase system can be represented as

$$U = u_1 V_1 + u_2 V_2 + u^* O \,. \tag{4.2}$$

The method of comparison with the ideal system is similarly used for the determination of all other extensive thermodynamic properties† which characterize the interface. For instance, for the entropy S of a two-phase system we obtain an expression analogous to (4.2):

$$S = s_1 V_1 + s_2 V_2 + s^* O, \tag{4.3}$$

where s_1 and s_2 are the entropy values per unit volume of each phase, and s^* is the specific excess surface entropy of the interface between 1 and 2.

With the aid of thermodynamic functions thus determined, and making use of the methods for the establishment of relationships between thermodynamic parameters, we obtain a system of correlations which takes into account the fact that interphase regions endowed with special properties exist in a heterogeneous system.

If the system under consideration is a multi-component one, the composition of region 3 in Fig. 21a differs from the composition of both phases in the bulk. Let us denote by n_i^* the number of moles of the i-th component, which would have to be added to the number of moles contained in the ideal system (without a transition region) in order to obtain the number of moles of the i-th component in the real system. We can then introduce a quantity

$$\Gamma_i = \frac{n_i^*}{O}, \tag{4.4}$$

which represents this excess related to the unit of interface. This magnitude is also called the quantity adsorbed per unit interface. It is obvious that this can be positive in case of positive adsorption, and negative in case of negative adsorption.

The quantities n_i^* and Γ_i depend on the position of the mathematical plane of separation in the ideal system. This indefiniteness disappears if this plane is always drawn in such a way that n_0^* and consequently Γ_0 are equal to zero for one component (generally the solvent).

The magnitudes which characterize the interface *i.e.* u^*, s^* and n_i^*, as well as other thermodynamic functions which can be derived from these

† Any property which, like mass, is characterized by the fact that its value for the entire system equals the sum of values for the individual phases of which the system is composed is called an *extensive property* or a *capacity factor*.

Any property which, like density, is constant throughout the entire phase so long as the phase is considered homogeneous by definition, and does not depend on the magnitude of the phase which is in a definite state, is called an *intensive property* or an *intensity factor*.

References p. 139–140

quantities make it possible to introduce the concept of a *surface phase* which corresponds to the transitional interphase region†.

Although this surface phase is not a classical bulk phase because it has zero volume, its energy, entropy and the amount of substance adsorbed in it, are proportional to the interfacial area in a manner exactly analogous to the three-dimensional case where the energy, entropy and mass are proportional to the volume. This method of describing the properties of heterogeneous systems is considerably more difficult in the cases of systems in which the degree of dispersion is so high that the linear dimensions of phases become commensurable with the thickness of the transition layer. This occurs for instance in thin films, *e.g.* foam films. Under such conditions the quantities characterizing the surface properties also become dependent on the linear dimensions of the phases (for films—on their thickness). These special cases require individual study and the question is treated further in Chapter VI.

The mechanical work dA^* required to increase the area of interface is proportional to the increase of this area:

$$dA^* = -\sigma \, dO. \tag{4.5}$$

The proportionality factor σ, known as "surface tension", plays an outstanding role in the physical chemistry of surfaces. This expression is analogous to the expression PdV for the work done during the expansion of bulk phases, dV corresponding to dO, and P to $-\sigma$. According to the above definition, σ is the work required to increase the interface area by 1 cm^2.

It follows from the above that, given sufficiently large phases, the heterogeneous system can be described by the parameters given in Table IV, 1.

The listed parameters of a system consisting of i components and two phases (1 and 2) include only one characteristic thermodynamic function, the entropy. This is sufficient for the calculation of the remaining characteristic functions (energy, enthalpy, free energy and thermodynamic po-

† The "surface phase" should be regarded not as a geometrically isolated region, but as some effective region characterized by excesses of thermodynamic magnitudes and introduced as a correction, the constancy of intensive factors being conserved for the bulk phases up to the idealized mathematical separation boundary.

Attempts at introducing greater geometrical definiteness and individualization for the surface require separate additional substantiation.

TABLE IV, 1

DESCRIPTION OF A HETEROGENOUS SYSTEM

For bulk phases	*For surface phases*	*For the heterogeneous system as a whole*
P	σ	P, σ
V	O	V, O
T	T	T
n_i	n_i^*	n_i, n_i^*
S	S^*	S, S^*

tential). It follows that in comparison with the description of a homogeneous system, two additional, qualitatively new parameters, σ and O are needed to describe a heterogeneous system.

To illustrate the method of establishing thermodynamic relations and functions by means of the above listed parameters, let us derive some of them.

For a constant temperature throughout the system, the variation of the total internal energy is given by the First Law of Thermodynamics

$$\mathrm{d}U = \mathrm{d}Q - \mathrm{d}A + \sum \mu_i \,\mathrm{d}n_i\,,$$

where μ_i is the so-called chemical potential of the i'th component. The heat brought into the system is

$$\mathrm{d}Q = \mathrm{T}\,\mathrm{d}S_1 + \mathrm{T}\,\mathrm{d}S_2 + \mathrm{T}\,\mathrm{d}S^*\,,$$

the mechanical work done by the system is

$$\mathrm{d}A = P_1\,\mathrm{d}V_1 + P_2\,\mathrm{d}V_2 - \sigma\,\mathrm{d}O\,,$$

and the non-mechanical work due to the change of composition is given by the sum $\sum \mu_i \,\mathrm{d}n_i$, taken over all the i components. This sum can be expanded for all phases

$$\sum \mu_i \,\mathrm{d}n_i = \sum_1 \mu_i \,\mathrm{d}n_i + \sum_2 \mu_i \,\mathrm{d}n_i + \sum_* \mu_i \,\mathrm{d}n_i\,,$$

so that we have correspondingly for the first (1), second (2) and surface (*) phases

$$dU = [T\,dS - P\,dV + \sum \mu_i\,dn_i]_1 + [T\,dS - P\,dV + \sum \mu_i\,dn_i]_2 + \\ + [T\,dS + \sigma\,dO + \sum \mu_i\,dn_i]_* . \qquad (4.7)$$

Expression (4.7) shows that, for a constant composition ($dn_i = 0$, $dn_i^* = 0$), constant volume ($dV = 0$) and constant surface area ($dO = 0$), we have

$$dU = dQ = T dS_1 + T dS_2 + T dS^* , \qquad (4.8)$$

i.e. the increase of internal energy is equal to the amount of heat supplied.

Similarly, making use of the definitive expression $F = U - TS$ for the free energy of the system, we have

$$F = U - T(S_1 + S_2 + S^*) ,$$

where $S_1 + S_2 + S^*$ is the total entropy. Taking into account that $d(TS) = T dS + S dT$ and making use of (4.7), we obtain

$$dF = [-S\,dT - P\,dV + \sum \mu_i\,dn_i]_{1,2} + [-S\,dT + \sigma\,dO + \sum \mu_i\,dn_i]_* . \qquad (4.9)$$

At constant temperature ($dT = 0$), volume ($dV = 0$) and composition ($dn_i = 0$, $dn_i^* = 0$) we have

$$dF = \sigma\,dO . \qquad (4.10)$$

Thus $(\partial F/\partial O)_{T,V,n_i} = f^*$, *i.e.* the free energy per unit surface area, is equal to σ when the adsorption is invariant ($dn_i^* = 0$) *i.e.* when the composition in the interface is independent of the interfacial area.

For the thermodynamic potential $\Phi = F + PV$ we obtain similarly, taking into account that $d(PV) = P dV + V dP$,

$$d\Phi = [-S\,dT + V\,dP + \sum \mu_i\,dn_i]_{1,2} + [-S\,dT + \sigma\,dO + \sum \mu_i\,dn_i]_* . \qquad (4.11)$$

Given a constant pressure, temperature and composition (which is always possible in the absence of adsorption ($dn_i^* = 0$), so that the components are not redistributed when the volume and surface are changed) we have

$$d\Phi = \sigma\,dO \quad \text{and} \quad \left(\frac{\partial \Phi}{\partial O}\right)_{P,T,n_i} = \sigma$$

Eqns. (4.7), (4.9) and (4.11) allow the thermodynamic parameters to be expressed by the derivatives of the characteristic functions. Let us recall this on the example (4.11) for a single bulk phase.

By definition $\Phi = \Phi(T, P, n_i)$, so that

$$\mathrm{d}\Phi = \left(\frac{\partial \Phi}{\partial T}\right)_{P,n_i} \mathrm{d}T + \left(\frac{\partial \Phi}{\partial P}\right)_{T,n_i} \mathrm{d}P + \sum \left(\frac{\partial \Phi}{\partial n_i}\right)_{T,P,n_j} \mathrm{d}n_i\,;$$

the subscript "j" in the last term indicates that all magnitudes of the n_i type, except n_i itself, are constant. On the other hand, according to (4.11)

$$\mathrm{d}\Phi = -S\mathrm{d}T + V\mathrm{d}P + \sum \mu_i \mathrm{d}n_i\,.$$

Since $\mathrm{d}T$, $\mathrm{d}P$ and $\mathrm{d}n_i$ are independent variables, it follows from the comparison of the two last expressions that

$$\left(\frac{\partial \Phi}{\partial T}\right)_{P,n_i} = -S\,, \quad \left(\frac{\partial \Phi}{\partial P}\right)_{T,n_i} = V \quad \text{and} \quad \left(\frac{\partial \Phi}{\partial n_i}\right)_{P,T,n_j} = \mu_i\,.$$

Making use of the rule of independence of the order of differentiation, *i.e.* that for $f = f(x, y)$

$$\frac{\partial}{\partial y}\left(\frac{\partial f}{\partial x}\right) = \frac{\partial}{\partial x}\left(\frac{\partial f}{\partial y}\right)$$

we obtain

$$\left[\frac{\partial}{\partial P}\left(\frac{\partial \Phi}{\partial n_i}\right)_{P,T,n_j}\right]_{T,n_i,n_j} = \left[\frac{\partial}{\partial n_i}\left(\frac{\partial \Phi}{\partial P}\right)_{T,n_i,n_j}\right]_{P,T,n_j}$$

or

$$\left(\frac{\partial \mu_i}{\partial P}\right)_{T,n_i,n_j} = \left(\frac{\partial V}{\partial n_i}\right)_{P,T,n_j}\,;$$

this yields

$$\left(\frac{\partial \mu_i}{\partial P}\right)_{T,n_i,n_j} = \bar{v}_i\,.$$

Hence we obtain for an incompressible liquid, after integration,

$$\Delta\mu_i = \bar{v}_i \Delta P\,.$$

If the system consists of only one component, we have correspondingly

$$\Delta\mu = \bar{v}\Delta P\,, \tag{4.12}$$

where $\bar{v}$ is the molar volume.

When applied to the ideal gas ($p\bar{v} = RT$), the same formula yields the following expression for the change of its chemical potential when the pressure is changed from p_1 to p_2 at a constant temperature:

$$\Delta\mu = RT \ln \frac{p_2}{p_1}. \tag{4.13}$$

Temperature dependence of surface tension

For the free energy of a system consisting of two phases and one component we can write an expression analogous to (4.2) and (4.1):

$$F = {}_1V_1 + f_2V_2 + f^*O, \tag{4.14}$$

where f^* is the free surface energy for unit surface area at constant temperature and volume. It has been seen that this is equal to the surface tension.

From this expression and from the known thermodynamic relationships it is possible to derive the temperature coefficient of surface tension for a one-component system. Let us consider a slight change $\mathrm{d}T$ in a system at temperature T, at constant volume, composition and surface area. According to (4.9), $\mathrm{d}F = -S\mathrm{d}T$. By differentiating (4.14) with respect to T, we obtain for the entire system

$$\left(\frac{\partial F}{\partial T}\right)_{V,O} = \left(\frac{\partial f}{\partial T}\right)_{V_1,O} V_1 + \left(\frac{\partial f}{\partial T}\right)_{V_2,O} V_2 + \left(\frac{\partial f^*}{\partial T}\right)_{V_1,V_2,O} O$$

or, using (4.3)

$$-S = -s_1V_1 - s_2V_2 - s^*O.$$

Therefore, for the surface phase we have

$$\left(\frac{\partial f^*}{\partial T}\right)_{V,O} \equiv \frac{\partial \sigma}{\partial T} = -s^*,$$

and since by definition $f^* = u^* - Ts^*$,

$$f^* \equiv \sigma = u^* + T\frac{\partial \sigma}{\partial T} \tag{4.15}$$

Experimental investigations of the dependence of surface tension on temperature indicate that, for temperatures not very close to the critical temperature, the surface tension decreases linearly with increasing temperature; according to (4.15) this means that in the given temperature interval the specific energy (u^*) is independent of temperature. This conclusion is completely explicable as it is known that the intermolecular forces which determine u^* depend only slightly on temperature. Furthermore, investigation of the temperature dependence of σ indicates that the temperature coefficient of surface tension $d\sigma/dT$ is small†, and therefore in the first approximation the term $T(\partial\sigma/\partial T)$ can be disregarded as compared with u^*. This makes it possible to evaluate σ on the strength of intermolecular forces acting in the liquid.

Capillary pressure

An important phenomenon in the chemistry of colloids is the increased pressure in a liquid possessing a convex surface. A convex surface is also associated with an increase in the vapour pressure of the liquid. Such convexity, with high curvature, appears when a phase is dispersed, and the associated special properties are therefore of fundamental importance for the study of colloids.

Let us assume that ΔO in Fig. 22 is a sufficiently small section of the curved interface between two phases whose volumes are V_1 and V_2, so that both main radii of curvature r_1 and r_2, which describe this section, are the same for all its points. The temperature T and the volume $V = V_1 + V_2$ are constant. When ΔO is displaced by dx towards the normal x (dotted line in Fig. 22) at this point, one of the volumes (*e.g.* V_1) will increase by dV_1, whereas the other will decrease by dV_2, so that $dV_1 = -dV_2$. On the other hand, geometrical considerations yield the following relations:

$$dV_1 = \Delta O\, dx \quad \text{and} \quad d\Delta O = \Delta O \left(\frac{1}{r_1} + \frac{1}{r_2}\right) dx\,.$$

Since the equilibrium condition is that $dF = 0$, according to (4.9), we have

$$dF = -\,P_1\, dV_1 - P_2\, dV_2 + \sigma\, d\Delta O = 0$$

(at T = const. and for a one-component system), where P_1 and P_2 are pressures in phases 1 and 2, respectively. Substituting $-dV_1$ for dV_2 into this expression and the expression obtained above for $d\Delta O$, we obtain

† Not too close to the critical temperature. Near the critical temperature the entropy term $T(\partial\sigma/\partial T) = -s^*T$ strongly increases, and the curve $\sigma = \sigma(T)$ touches the abscissa near the critical point, *i.e.* at the critical temperature σ becomes equal to zero, and $T(\partial\sigma/\partial T)$ becomes equal to u^*.

References p. 139–140

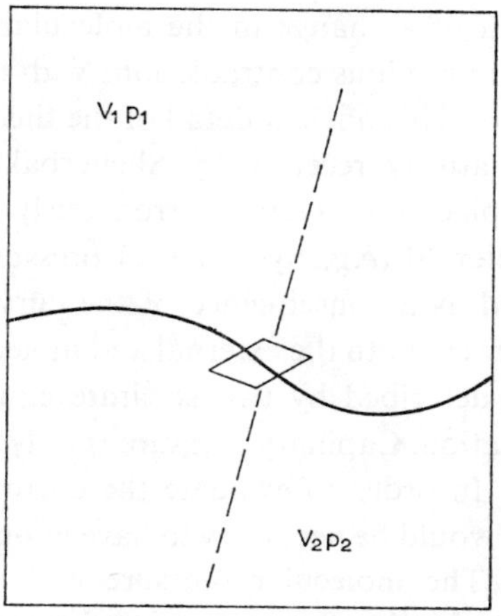

Fig. 22. Curved interface between two phases. The parallelogram represents a small area ΔO of the interface, the normal to it being indicated by the broken line.

$$P_1 - P_2 = \sigma\left(\frac{1}{r_1} + \frac{1}{r_2}\right) \tag{4.16}$$

which, in the case of a spherical surface element ($r_1 = r_2 = r$) yields

$$P_1 - P_2 = 2\sigma/r. \tag{4.16'}$$

Expressions (4.16) and (4.16′) determine the so-called capillary pressure $\Delta P = P_1 - P_2$. Eqn. (4.16) allows the calculation of the pressure drop ΔP associated with the crossing of any surface; for this purpose it is necessary to know the characteristic radii of its curvature r_1 and r_2 at various points. It is clear from the above equations that in the case of a plane surface ($r_1 = r_2 = \infty$) the capillary pressure is equal to zero* and the pressure is identical in the entire system, this being the condition of equilibrium for all cases when the interphase regions are not taken into account. Expressions (4.16) and (4.16′) show also that ΔP can have positive and negative values. We shall see below that these formulae serve as a basis of surface tension measurements.

Eqn. (4.16) was first derived in 1806 by Laplace[5], by a somewhat different approach. In this derivation the capillary pressure ΔP was

* Capillary pressure is also cancelled, according to (4.16), at the "inflexion points" where $r_1 = -r_2$.

regarded as the result of a change in the molecular pressure within the liquid, which results in obvious contradictions with respect to the sign of the molecular pressure. This unclear detail of the theory of capillarity was finally elucidated relatively recently by Shcherbakov (1958)[6], who demonstrated that Laplace's derivation erroneously identifies molecular pressure with the external (*e.g.* hydrostatic) pressure. In fact the equilibrium state attained as a consequence of the curvature of the surface corresponds to changes in both the external and molecular pressures, these two pressures being described by two separate equations analogous to that of Laplace's equation. Capillary pressure is only related to the change of external pressure. In order to evaluate the corresponding changes of molecular pressure it would be necessary to have at one's disposal methods of its measurement. The molecular pressure within a liquid, which is determined by intermolecular forces and is of great importance for the molecular-kinetic theory of liquids, cannot be investigated by capillarity measurements. It will be shown below that this is only possible if we study the properties of microheterogeneous systems, such as very thin liquid films.

Thomson–Gibbs equation

Let us consider a one-component system consisting of two phases, for instance, a liquid 1 and its vapour 2. The pressure change in the liquid which occurs when its surface is curved is accompanied by a change of vapour pressure. This is demonstrated below.

At equilibrium the chemical potentials of both phases are identical*:

$$\mu_1 = \mu_2 \quad \text{and} \quad \Delta\mu_1 = \Delta\mu_2 \,,$$

where $\Delta\mu$ represents the variation in μ due to a variation in curvature, and $\Delta\mu_1$ and $\Delta\mu_2$ are calculated with respect to the same initial state of both phases, *e.g.* μ_∞ for a liquid with a large plane surface in equilibrium with its vapour.

* This condition is demonstrated *e.g.* in the following manners: when dn_1 moles of substance are transferred from phase 1 to phase 2 at constant temperature and pressure we have according to (4.11), $d\Phi = \mu_1\, dn_1 + \mu_2\, dn_2$. For the entire system $dn = dn_1 + dn_2 = 0$. Since at equilibrium $d\Phi = 0$, $d\Phi = (\mu_1 - \mu_2)\, dn_1 = 0$, whence $\mu_1 = \mu_2$.

$\bar{v}$ = molar volume

According to (4.12), for an incompressible liquid

$$\Delta\mu_1 = \bar{v}\Delta P\,,$$

and, according to (4.13), for an ideal gas

$$\Delta\mu_2 = RT\ln\frac{p_r}{p_\infty}$$

where p_r is the vapour pressure over a curved surface of radius r, and p_∞ is the vapour pressure over a plane surface ($r = \infty$). We then obtain directly the Thomson–Gibbs or Kelvin equation:

$$\Delta P = \frac{2\sigma}{r} = \frac{RT}{\bar{v}}\ln\frac{p_r}{p_\infty}\,. \tag{4.17}$$

Consequently, the vapour pressure over a small spherical surface of a liquid,

$$p_r = p_\infty \exp\left(\frac{2\sigma\bar{v}}{RTr}\right)$$

will be greater than the vapour pressure over a plane surface of a large mass of the same liquid; in other words the smaller drops (or crystals) should possess a higher vapour pressure than the larger ones. It will be shown (p. 208) that this is the reason why substance is transferred from finer particles to the coarser ones so that the fine particles become even finer, and the coarse ones grow at their expense until the first fraction has completely disappeared. It can be demonstrated that the same considerations are applicable to solubility. For an adequate assessment of the practical effect, it is necessary to take into account that, firstly, the transfer of substance from certain particles to others occurs by diffusion, which is a slow process, especially in solution and that, secondly, $\ln(p_r/p_\infty)$ attains measurable values only in the case of very small particles.

Adsorption and surface tension

A thermodynamic method of defining adsorption has already been discussed. Let us now determine the relationship between the concentration c of a solute in the interior of the phase (bulk concentration) and its excess concentration Γ on the surface, for the relatively simple case of a two-component system.

When a small quantity dn^* of a substance is adsorbed at constant temperature ($\mathrm{d}T = 0$), the increase of the free surface energy is, according to (4.9),

$$\mathrm{d}F^* = \sigma\,\mathrm{d}O + \mu^*\,\mathrm{d}n^*\,.$$

On the other hand, since $F^* = \sigma O + n^*\mu^*$,

$$\mathrm{d}F^* = \sigma\,\mathrm{d}O + \mu^*\,\mathrm{d}n^* + O\,\mathrm{d}\sigma + n^*\,\mathrm{d}\mu^*\,;$$

by equating these two expressions we obtain $O\,\mathrm{d}\sigma + n^*\,\mathrm{d}\mu^* = 0$.

Hence for $\Gamma = n^*/O$ (at equilibrium $\mathrm{d}\mu^* = \mathrm{d}\mu$)

$$\Gamma = -\left(\frac{\partial\sigma}{\partial\mu}\right)_T \tag{4.18}$$

or, in the special case of sufficiently dilute solutions, when $\mathrm{d}\mu = (RT\,\mathrm{d}c)/c$

$$\Gamma = -\frac{c}{RT}\left(\frac{\partial\sigma}{\partial c}\right)_T, \tag{4.18'}$$

These equations are called the Gibbs adsorption isotherm.

Eqns. (4.18) and (4.18′) are fundamental in all adsorption processes.

Equilibrium at the boundary of three phases. Contact angle

If a drop applied on to the surface of another substance does not spread completely, then along its periphery three phases are simultaneously in contact: liquid drop, vapour and solid base. The angles formed by the three interfaces are obviously not arbitrary. If the contact perimeter changes while the amount of the liquid remains constant, the drop changes its shape and the drop/vapour, drop/base and base/vapour interfaces are also affected, bringing about a change in the total surface energy of the system related to the areas of these three interfaces. Equilibrium is achieved when this surface energy is at its minimum, this condition being associated with certain angles of "incidence" of the three phases. To determine these angles one has to equate the variation of the total free surface energy at a constant drop volume (and constant temperature) to zero, as this corresponds to the condition of the minimum total energy of the system. A relatively simple solution can be obtained if the case of a solid base is analysed, assuming that the drop is a spherical

References p. 139–140

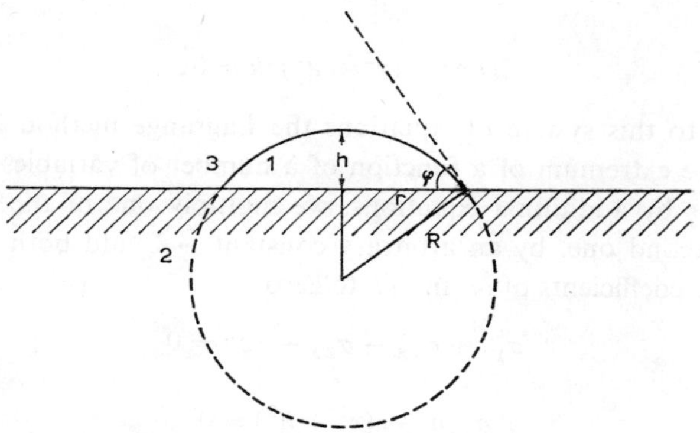

Fig. 23. Derivation of the equilibrium contact angle.

segment. This corresponds to the shape of a drop with the minimum drop/vapour surface* (Fig. 23).

Making use of known geometrical formulae, we find that the free surface energy of the system will be

$$F^* = \sigma_{23}O_{23} + \sigma_{12}O_{12} + \sigma_{13}O_{13} = (O - \pi r^2)\sigma_{23} + \pi r^2 \sigma_{12} + \pi(r^2 + h^2)\sigma_{13} ;$$

the constant volume of the drop will be

$$V = \tfrac{1}{6}\pi h(3r^2 + h^2) = \text{const.},$$

and

$$\cos\varphi = \frac{r^2 - h^2}{r^2 + h^2},$$

where φ is the *contact angle*, and O is the surface area of the base, hence for a variation in r:

$$\delta F^* = -\sigma_{23} 2\pi r \delta r + \sigma_{12} 2\pi r \delta r + \sigma_{13}\pi(2r\delta r + 2h\delta h) = 0$$

and

$$\delta V = \frac{\pi}{6}(6rh\delta r + 3h^2\delta h + 3r^2\delta h) = 0 ;$$

or,

$$r\delta r(\sigma_{12} + \sigma_{13} - \sigma_{23}) + h\sigma_{13}\delta h = 0$$

* If the force of gravity is also taken into account then the condition of the energetically preferred drop shape becomes more complex; it is also necessary to allow for its potential energy in the external field.

and

$$2rh\delta r + (r^2 + h^2)\delta h = 0.$$

Applying to this system of equations the Lagrange method for finding the relative extremum of a function of a number of variables related to each other by additional equations, we multiply one of the equations, *e.g.* the second one, by an arbitrary constant $-\lambda$, add both equations, equate the coefficients of δr and δh to zero:

$$\sigma_{12} + \sigma_{13} - \sigma_{23} - \lambda 2h = 0$$

and

$$\sigma_{13}h - \lambda(r^2 + h^2) = 0$$

and solve for r and h:

$$r^2 = \frac{2\sigma_{13}(\sigma_{12} + \sigma_{13} - \sigma_{23}) - (\sigma_{12} + \sigma_{13} - \sigma_{23})^2}{4\lambda^2}$$

$$h^2 = \frac{(\sigma_{12} + \sigma_{13} - \sigma_{23})^2}{4\lambda^2}.$$

Finally, after substituting r^2 and h^2 into the expression for $\cos\varphi$ we obtain

$$\cos\varphi = \frac{\sigma_{23} - \sigma_{12}}{\sigma_{13}}. \tag{4.19}$$

Eqn. (4.19) is usually derived by a much simpler method, by making use of the concept that σ is a force per unit length, which acts along the surface. In this case the equilibrium contact angle corresponds to the equilibrium of forces on the periphery of the drop, *i.e.* σ_{23} is equated to σ_{12} plus the component along the surface, $\sigma_{12}\cos\varphi$. However, this interpretation of surface tension is arbitrary, and it will later be shown that there exists no tangential force along the surface of a pure liquid.

The above derivation makes it possible to calculate any more complicated case, *e.g.* a case involving the effect of gravity with the corresponding deformation of the drop, as well as a special case of a very small droplet in equilibrium with a thin polymolecular equilibrium layer which wets the substrate.

Microheterogeneous systems

The ideas presented so far are not sufficient for the explanation of all

References p. 139–140

thermodynamic properties of disperse systems. Their inadequacy is shown particularly strongly in the transition region of dispersion, where a gradual transition from heterogeneous disperse systems to molecularly disperse systems should be observed. In accordance with the previously given thermodynamic concepts, the total interface and simultaneously the total excess surface energy, increases with increasing degree of dispersion, and consequently with increasing number of particles and reduction of their size. This means that the higher the degree of dispersion of a system the further it is removed from the state of equilibrium, and therefore the less stable it is. On the other hand, it is known that the maximum (molecular) degree of dispersion corresponds to thermodynamically stable states. The gap observed in the sequence of thermodynamic analysis comprises the high-molecular substances, and therefore deserves close attention. The recent attempts at overcoming this difficulty have not yet been completed and integrated. A brief description of the problem will however be given in view of its importance.

The shortcomings of the above thermodynamics of heterogeneous systems are summarized in two points. In the first place no allowance is made for fluctuations whose role increases, as will be demonstrated, with an increasing degree of dispersion. Secondly, as the degree of dispersion is increased, particle size is reduced and curvature of their surfaces is increased, the applicability of the considerable simplification introduced by Gibbs with regard to the independence of thermodynamic parameters which characterise the phase surface, and of the phase size and the curvature of its surface, becomes less and less justified. Thus *e.g.* the question of establishing what are the smallest particle sizes at which one can still envisage the presence of the phase surface and of the corresponding surface tension loses any significance, and there arises the question in what way the magnitude of surface tension varies gradually when very small particles become even smaller. It is obvious that, in view of the discrete molecular structure of the substance, a sudden change of properties cannot be avoided on transition to the molecular degree of dispersion. Nevertheless it is in this way possible to approach this limit more closely.

Let us consider both features referred to above in greater detail.

As has already been mentioned in the preceding chapter, spontaneous deviations from the equilibrium state (fluctuations) are more probable if fewer molecules are involved. Moreover, any parameter can fluctuate, and the probability of such fluctuation is determined solely by the resulting change of the characteristic function of state, *e.g.* the free energy, as compared with its value at equilibrium. In 1913, Mandelstamm[7] first applied these ideas to the interfacial area by analogy with the theory of Brownian movement assuming that waves can arise in the liquid interface (liquid/liquid or liquid/gas) as a result of molecular motion, *i.e.* that such a surface can increase spontaneously by fluctuations. No detailed analysis of such "Brownian" waves on the surface of the liquid (carried out in the paper referred to above) will be given in this book, particularly as this phenomenon has not as yet been studied quantitatively. It is, however, possible to estimate the amplitude of these waves on the basis of elementary ideas, using the frequently applied expression for mean-square fluctuations (3.18)

$$\overline{(\Delta\lambda)^2} = \frac{kT}{\partial^2 F/\partial\lambda^2}.$$

In the present case the fluctuation consists of the generation of a wave with an amplitude h on the surface of the liquid. For sufficiently small h the increase of surface area as compared with the plane surface is equal to πh^2, and is little affected by the shape of the wave, whereas the corresponding increase of free energy is $\Delta F = \pi h^2 \sigma$. The gravitational component which, for small h is infinitely small (as is known from the theory of surface waves) is of course neglected. In this case $\partial^2 F/\partial\lambda^2 = \partial^2 F/\partial h^2 = 2\pi\sigma$ and

$$\sqrt{\overline{(\Delta h)^2}} = \sqrt{\frac{kT}{2\pi\sigma}}$$

The above equation shows that at room temperature ($kT \approx 4 \cdot 10^{-14}$), the amplitude of a spontaneous surface wave $\sqrt{\overline{(\Delta h)^2}}$ is greater by one order of magnitude than the molecular dimensions ($\sqrt{\overline{(\Delta h)^2}} = 10$ Å) at $\sigma \approx 1$, *i.e.* at small values of surface tension. As is known from the temperature dependence of surface tension, the latter becomes zero at the critical temperature, and therefore near this temperature can be maintained at a very low level. Mandelstamm verified the existence of the presumed effect by observing the diffusion of light by an obliquely illuminated interface between two partially miscible liquids. The increase of light scattering at temperatures close to the critical solution temperature of both liquids, when σ is very small at the interface, could be clearly discerned.

The spontaneous increase of surface due to fluctuations can go even further, up to the severance of particles from the surface, and even to the complete comminution of one phase within another, to form a large number of small particles. At the critical temperature, when σ is almost zero, there arises such a state of self-dispersion with a definite equilibrium size distribution of the particles. Since systems of this kind are capable of scattering light intensely, this phenomenon is sometimes called the critical opalescence. The theory of heterophase fluctuations, developed by Volmer[8], gives corresponding formulae for the size distribution of such particles. Recently (1959) Volmer completed the theory of critical opalescence, having described on the basis of this distribution the produced light scattering.

The phenomenon of critical opalescence and the theory of heterophase fluctuations indicate the basic possibility of spontaneous dispersion, resulting in the formation of thermodynamically stable heterogeneous disperse systems. These effects and systems are, however, of a highly special nature, and cannot therefore form the basis for a general treatment of colloids. Quite recently Rebinder and co-workers have been working on the development of a more general case, the increase of thermodynamic stability of disperse systems due to the fluctuation factor when the surface tension is lowered by surface-active substances. This approach is very tempting since it is known that surface-active substances usually behave as stabilizers in colloidal solutions. From this standpoint it is attempted to characterize and explain the "lyophilic nature" as a result of the decrease, in the given system, of the free surface energy by adsorption. It can be seen from the aforesaid that the fluctuation effect begins to play a marked role when the surface tension becomes very small. Confirmation of these interesting ideas makes it therefore necessary to measure very small surface tensions; this is accompanied by special difficulties. Recently (1958) these ideas have been also extended to the phenomena of the lowering of strength and to the disintegration of solids owing to the effect of surface-active substances. In a remarkable experiment Rehbinder[9] disintegrated a zinc monocrystal to powder by applying to it minute quantities of gallium, which is a surface active agent in this case. In 1958 Shchukin[10] gave a theoretical interpretation of this phenomenon, assuming that gallium penetrates into the microcracks of the mosaic structure of the zinc monocrystal and lowers its surface tension to a very low level, thus enabling the fluctuation-induced weakening of bonds between the blocks of the micro-crystalline structure to be produced. Although this explanation has not as yet been proved and although the treatment of the effect appears to be simplified, not taking into account the fact that the mosaic has a system of micro-fissures which serve as "nuclei of disintegration", or that the peculiar properties of the thin layers of the zinc/gallium solution, forming in these microfissures, can also play a dispersing role, the development of the fluctuation theory of colloid stability is undoubtedly of fundamental importance.

The probable dispersing effect of thin layers of solution in the micro-fissures of the mosaic would be a manifestation of the second of the two above-mentioned peculiar features of the microheterogeneous state which we are now going to analyse.

Gibbs assumed that the transition layer in the interface region, in which the gradual change of properties takes place, is very thin. He was, however, unable to establish the exact value of this magnitude, and thus to determine the lower limit of phase dimensions at which it would still be possible to apply the concept of a surface phase having parameters which are independent of the thickness, because no data regarding molecular size or the range of intermolecular forces were available at that time. In spite of the above Gibbs assumed the existence of such a limit. Thus *e.g.* when analysing the problem of the stability of foam[4], he definitely speaks of the appearance of special properties in very thin layers, which can increase their instability and cause their breakdown. A clearer thesis dealing with the change of thermodynamic properties (chemical potential) in thin layers was put forward by Polanyi in 1914[11]. The latter author postulated that, as a result of interactions between molecules in the thin polymolecular layer and the base on which this layer is supported, there may take place such a change of the chemical potential of the layer substance as to cause the lowering of its equilibrium vapour pressure; a layer of a certain thickness will thus be in thermodynamic equilibrium with an unsaturated atmosphere. The resultant situation corresponds to polymolecular adsorption and forms its theoretical substantiation. Although, as will be demonstrated in the special analysis of thin layers, Polanyi's theory of polymolecular adsorption is incomplete, it is fundamentally correct. Nevertheless, it encountered serious objections. Starting from the idea that intermolecular attraction forces vary inversely with the seventh power of distance, Hückel[12] considered that the influence of the base on the polymolecular adsorption layer cannot be expected to extend further than one or two molecular diameters. The lack of substantiation

References p. 139–140

of Hückel's criticism becomes clear from consideration of Hamaker's work in 1937[13], who calculated the energy of the interaction of two phases across the separating gap as the sum (integral) of the interaction, of all molecules in one phase with all molecules in the other. It was found that, as a result of summation, the energy of interaction of the two phases varies inversely not as the sixth power of distance (as in the case of separate molecules), but as the square of distance, *i.e.* much more slowly. It was thus shown that the distances from the phase surface, at which the influence of the second phase can still be perceived, are much greater than the dimensions of molecules.

It will be shown further, in the analysis of the electrokinetic phenomena, that under certain conditions, a so-called diffuse double layer forms at the interface in aqueous solutions of electrolytes. The diffusion of excess charges (ions) present in this layer from the interface towards the bulk of the solution can extend to a depth of 1 μ at small electrolyte concentrations, so that the thickness of such diffuse layers may be of the order of 10^4 Å.

The above examples, which will be quantitatively discussed in a further section, are sufficient to show that the transition interphase layer 3 in Fig. 21 can have a thickness considerably greater than the molecular diameter. This in turn leads to the possibility that the thermodynamic properties of disperse systems deviate from Gibbs' theory at high degrees of dispersion, but still within the region where the particles of the disperse phase are much larger than molecules.

A theoretical analysis of the thermodynamic behaviour of systems in which, in view of the effect referred to above, the characteristic "surface" properties are not independent of the volume and shape of small phases, is extremely difficult and has not yet been carried out in a general form. On the other hand, the theory of thin plane-parallel layers of substances has been developed in greater detail. In this case, the curvature factor is no longer present and the layer thickness h is the only remaining factor which determines the dimensions of the microphase. To characterise the special thermodynamic states of thin layers Deryagin (1934)[14] introduced the concept of "disjoining pressure", *i.e.* the excess of layer pressure over the pressure of a large mass of the liquid; in the case of a one-component layer, this parameter is a measure of the excess chemical potential in the layer. The isotherm of this disjoining pressure, *i.e.* its dependence on the thickness of layer at constant temperature, was found to be a very convenient and directly measurable function of state of thin liquid layers. Other authors (Overbeek's school) prefer to use for this purpose the thickness-dependence of the excess energy of the thin layer per unit area. Both methods of approach are equivalent since this energy is unambiguously related to the disjoining pressure.

Deryagin and Martynov (1962)[15] have attempted to extend the thermodynamic theory of thin layers to multi-component systems.

Starting from 1958, Shcherbakov[16] has suggested a thermodynamic treatment for the case of very small isodiametric droplets and crystals; this introduces a correction into Thomson–Gibbs' equation (4.17) for the dependence of the vapour pressure of such small formations on their linear dimensions. The same author (1959–1961)[17] has also analysed the heat of sublimation of small crystals and the heat of evaporation of small droplets. In 1960 Shcherbakov and Ryazantsev[18] analysed the interesting case, observed in 1938 by Frumkin[19], of the equilibrium of a drop lying on a surface wetted with a polymolecular layer of the same liquid as the drop; this is a development of the contact angle theory*.

The importance of all these thermodynamic investigations and the field of their application are ultimately determined by the magnitude of molecular constants which can be either calculated on the basis of certain models or determined from the experimental study of the analysed effects, *e.g.* the isotherm of the disjoining pressure of a thin layer. A more detailed discussion of these problems will therefore be carried out (Chap. VI) during the special analysis of thin liquid layers, as the fundamental importance of this field justifies in the author's opinion a special chapter devoted to this topic.

2. Formation of New Phases

The formation of a new phase in a supersaturated system is a particularly kinetic problem. The kinetics of this process—the rate of formation of the new phase—depend, however, substantially on the magnitude of a certain energy barrier, *i.e.* on the so-called work of formation of the nucleus of the new phase. This work, which may be calculated by a purely

* A number of authors [starting with Tolman (1949)[20], and later König, Buff[21], Hill[22] and others] carried out analogous theoretical studies on the effect of curvature of the phase surface and of the phase dimensions on thermodynamic parameters, making use of the geometrically defined idea of a surface phase, which presents a much more limited treatment of the problem.

thermodynamic method as was shown for the first time by Gibbs[4], is equal to $\frac{1}{3}$ of the free surface energy of a drop having such dimensions that its vapour pressure is the same as that of the supersaturated system. Therefore, if we confine the discussion to analysing only the energy barrier referred to above, it is logical to relate this problem to the thermodynamics of heterogeneous systems.

It was already indicated in Chapter I that disperse systems are formed either by the dispersion of coarser phases or by condensation from molecules in supersaturated solutions. The production of new phases by condensation is therefore a most important fundamental problem in colloid science. Unfortunately, the theory of this process and its experimental investigation are far from complete. The clearest results have now been obtained in the simplest cases, in which the new phase forms in a gaseous medium, *i.e.* for the formation of aerosols. In this elementary exposition of the problem we shall thus confine the account to examples taken from the field of aerosols.

Formation of a new phase by condensation is characterised by the rate of formation, equal to the number of particles forming in unit time in a unit of volume. If particles of a given type (*e.g.* water drops from water vapour) can be formed in more than one way, the rate I of this process will be the sum of the component partial rates, $I_0, I_1 \ldots$, so that

$$I = I_0 + I_1 + \ldots \qquad (4.20)$$

It is usually found under a given set of conditions that one of the partial rates is considerably greater than the others, and even than their sum, and therefore in any real case it is possible to consider only one mechanism. It is therefore important to explain the factors which determine magnitudes of I_0, I_1, etc.

In the first place let us consider the simplest case of phase formation by condensation of aerosol particles from a supersaturated space free from condensation centres—the so-called homogeneous generation of a new phase in a supersaturated system. The theory of this process, developed by Volmer[23], forms the basis of all more complex cases which will be analysed below.

The basic factor which determines the formation of a new phase is the supersaturation. It is only in exceptional cases, when the system is close to the critical state, that a new phase can form in the near absence of supersaturation, since the very small particles condensing in the first place have a highly convex surface, and their vapour pressure is consequently higher than the vapour pressure of a large mass (plane surface) of the same substance. The vapour pressure p_r of a surface of radius r, corresponding to a droplet of the same radius, is expressed by the Thomson–Gibbs equation (see p. 89)

$$p_r = p_\infty \exp\left(\frac{2\sigma\bar{v}}{rRT}\right)$$

where p_∞ is the vapour pressure of the plane surface of the corresponding liquid.

A small liquid drop in an atmosphere of its vapour at a pressure p_∞ (equilibrium of a large phase) will thus have a vapour pressure p_r higher than that in the ambient medium. The drop can therefore be in a state of (unstable) equilibrium only with vapours which are supersaturated with respect to the bulk phase, and every degree of supersaturation will correspond to an equilibrium droplet of a strictly defined radius—the so-called nucleus of the new phase. For such equilibrium drops to form in the homogeneous vapour of a given supersaturation, they must be developed from smaller drops the vapour pressure of which is greater than the vapour pressure in the system. In the case of a sufficient degree of supersaturation these equilibrium critical drops are sufficiently small and can form as a result of fluctuations in density and concentration.

The process of phase formation is thus of the fluctuation type so that the number of the resultant nuclei, and therefore the required rate of formation of the new phase I_0, is determined by the number of corresponding fluctuations. More precisely, I_0 is proportional to the probability of the occurrence of these fluctuations, which is known to be proportional to $\exp(-\Delta F/kT)$ where ΔF is the change of free energy during the given fluctuation. In the case of an isothermal and reversible formation of a nucleus ΔF is equal to the corresponding work A_{k_0} so that, finally,

$$I_0 = k_0 \exp\left(-\frac{A_{k0}}{kT}\right). \tag{4.21}$$

It follows from the above that I_0 is determined in principle by the work A_{k_0} required for the formation of a nucleus of the new phase, *i.e.* a molecular aggregate whose vapour pressure is equal to the increased vapour pressure of the supersaturated system.

I_0 can be calculated by statistical methods [Farkas (1927)[24], Stranski

and Kaishev (1934)[25], Becker and Döring (1935)[26], Frenkel (1939)[27], Zel'dovich (1943)[28] and others], and both constants k_0 and A_{k_0} can be determined. However, if only the energy barrier A_{k_0} is to be calculated, which is very important in this case, this can be accomplished, as already mentioned, by purely thermodynamic methods [Volmer and Weber (1925)[29]]. The work of the isothermal formation of a nucleus is equal to the change taking place in the free energy of the system when the nucleus (drop) is formed within it. If the vapour (n_k moles) had an initial free energy $F_1 = n_k\mu_1$, then after the formation of the critical drop the free energy is $F_2 = n_k\mu_2 + \sigma O_k$.

The chemical potential of the liquid (μ_2) in the drop differs from the chemical potential in the vapour (μ_1), since in the drop the liquid is exposed to a capillary pressure of $2\sigma/r_k$, where r_k is the drop radius; the change of the chemical potential will therefore be

$$\mu_2 - \mu_1 = -\frac{2\sigma}{r_k}\bar{v}.$$

The work of nucleus formation is obtained by adding to the surface energy of the drop the change of free energy caused by the change in the drop's chemical potential. Consequently:

$$A_{k_0} = 4\pi r_k^2\sigma - \frac{2\sigma}{r_k}\bar{v}\frac{4\pi r_k^3}{3\bar{v}} = \frac{4\pi r_k^2\sigma}{3} = \frac{\sigma O_k}{3}, \tag{4.22}$$

i.e. the work of the isothermal and reversible nucleus formation is equal to $\frac{1}{3}$ of the total surface energy of the nucleus. This result, which, as mentioned on p. 99 had been noted by Gibbs, enables A_{k_0} to be calculated as a function of supersaturation, making use of (4.17) for the calculation of r_k, this yields

$$A_{k_0} = \frac{16\pi\sigma^3\bar{v}^2}{3R^2T^2\ln^2\dfrac{p_r}{p_\infty}} \tag{4.23}$$

where the supersaturation is determined as $\ln(p_r/p_\infty)$. Substituting A_{k_0} into (4.21), we obtain

$$I_0 = k_0\exp\left(-16\pi\sigma^3\bar{v}^2N_A/3R^3T^3\ln^2\frac{p_r}{p_\infty}\right). \tag{4.24}$$

The kinetic constant k_0, calculated by the statistical method of Becker and Döring (1935)[26], is proportional to the concentration of the vapour from which the new phase is obtained. Without going into the details of the statistical method of evaluation, it can be said that this relationship is understandable since the probability of becoming the centre of nucleus formation is the same for all molecules.

In 1934 Volmer and Flood[30] carried out an experimental study of the kinetics of homogeneous condensation of vapours in the absence of extraneous particles, and at the same time verified experimentally Volmer's theory referred to before. These studies are of great fundamental significance since, with their aid the theory of phase formation is verified under particularly simple conditions, which makes it possible to evaluate all the process-determining parameters. The investigations were carried out in a Wilson chamber in which, owing to a sufficient adiabatic expansion of air saturated with the vapour of the liquid under examination, these vapours are cooled and supersaturated giving rise to a cloud. The rate of nuclei formation is checked visually by determining the beginning of condensation, *i.e.* the minimum supersaturation at which the cloud appears. In these conditions I_0 is of the order of unity. Since I_0 depends exponentially on supersaturation $\ln(p_r/p_\infty)$, this approximate method of determination of the "beginning" of condensation is sufficiently precise to confirm the theory with regard to the value of the exponent which determines A_{k_0} and for the evaluation of the order of magnitude of the pre-exponential factor.

Without describing in any detail this and the following studies [Sander and Damkhöler (1943)[31], Clark and Rodebush (1953)[32]], it need only be noted that they bear out satisfactorily the theory. It is of course very essential for the system not to contain any previous condensation centres on which the formation of drops would occur much more easily. Such "purification" is easily carried out in the Wilson chamber by repeated condensation: all centres are precipitated from the gaseous phase, and the critical supersaturation corresponding to the beginning of the formation of the new phase is increased. When this critical supersaturation attains its maximum value and does not increase on further condensation, it can be assumed that complete "purification" of the system has been achieved and that phase formation takes place without any condensation centres.

Experimental results and calculations show that supersaturations (p_r/p_∞) required to attain measurable rates of nuclei formation without any extraneous condensation centres are very high—of the order of

References p. 139–140

500%. Such high degrees of supersaturation are usually not observed in natural conditions, *e.g.* in the atmosphere. We might therefore assume that the formation of the new phase is usually facilitated by the presence of centres which reduce the value of A_{k_0}. Such centres can be either ions or solid or liquid particles. Let us first consider the influence of ions.

The part played by ions in the condensation process in a gaseous medium is evident from Wilson chamber practice. It is known that the path of the products of radioactivity decay and of other particles endowed with high energy (*e.g.* cosmic ray particles) can be observed in this chamber, owing to the fact that condensation proceeds more easily on ions formed when these particles collide with molecules of the gas. A cloud track is formed on the ions which mark the path of the particle, rendering this path visible, even when the supersaturation is still insufficient for cloud formation in unionised gas.

The theory of phase formation in vapours (in the presence of charged particles) was developed by Thomson in 1906[33] and later by Tohmfor and Volmer in 1938[34]. To understand its physical meaning, let us imagine a liquid drop which is a conductor. When such a drop collides with an ion the charge of the latter will be distributed on the surface of the drop, resulting in a decrease of the surface tension and of the work of nucleus formation.

The lowering of surface tension as a result of the electrification of the surface can be observed directly in the so-called electrocapillary phenomena. If for example we polarize the surface of mercury by applying a potential to mercury held in a capillary in contact with a solution of an electrolyte* (Lippmann's electrometer), the mercury level will rise owing to the lowering of surface tension at the mercury/solution interface.

The relationship between the change of surface tension and the change of charge at the interface can be derived thermodynamically. For this purpose it is expedient to introduce the *electrochemical potential*, $\bar{\mu}$, of the charged component (*e.g.* of ions which are adsorbed on the interface). If we confine ourselves to the case when the component is found only at the interface, then the term $\bar{\mu}\,d\bar{n}$ (where $\bar{n}$ is the number of moles or gram-ions of the charged component) will be equal to the electrical work of charging a unit area when the latter adsorbs dn moles of the charged component which carry an electric charge dq. This electrical work can be

* Mercury in the capillary tube is connected to a mercury bath in a wide vessel.

obviously expressed as $\varphi\,\mathrm{d}q$, where φ is the electric potential with respect to the surface. Thus

$$\bar{\mu}\,\mathrm{d}\bar{n} = \varphi\,\mathrm{d}q\,. \tag{4.25}$$

The assumption that the electric component is concentrated on the surface means also that the variation of the free energy of the system $\mathrm{d}F$ can be expressed only by its variation in the surface phase. Consequently

$$\mathrm{d}F = \mathrm{d}F^* = -S^*\,\mathrm{d}T + \sigma\,\mathrm{d}O + \bar{\mu}\,\mathrm{d}\bar{n} = -S^*\,\mathrm{d}T + \sigma\,\mathrm{d}O + \varphi\,\mathrm{d}q\,; \tag{4.26}$$

furthermore

$$\mathrm{d}F^* = \left(\frac{\partial F^*}{\partial T}\right)_{O,q}\mathrm{d}T + \left(\frac{\partial F^*}{\partial O}\right)_{T,q}\mathrm{d}O + \left(\frac{\partial F}{\partial q}\right)_{T,O}\mathrm{d}q\,, \tag{4.27}$$

since, according to (4.26), $F^* = F^*(T,O,q)$. It follows from (4.26) and (4.27) that

$$\left(\frac{\partial F^*}{\partial T}\right)_{O,q} = -S^*\,,\quad \left(\frac{\partial F^*}{\partial O}\right)_{T,q} = \sigma \quad \text{and} \quad \left(\frac{\partial F^*}{\partial q}\right)_{T,O} = \varphi\,. \tag{4.28}$$

Since the sequence of differentiation is immaterial

$$\left[\frac{\partial}{\partial q}\left(\frac{\partial F^*}{\partial O}\right)_{T,q}\right]_{O,T} = \left[\frac{\partial}{\partial O}\left(\frac{\partial F^*}{\partial q}\right)_{T,O}\right]_{q,T}$$

or, according to (4.28),

$$\left(\frac{\partial \sigma}{\partial q}\right)_{O,T} = \left(\frac{\partial \varphi}{\partial O}\right)_{q,T}\,. \tag{4.29}$$

This is Lippmann's equation[35] which forms the basis of the explanation of electrocapillary curves*.

With the aid of this equation it is possible to calculate the variation of surface tension during the charging of the drop. According to (4.29), for given O and T we have:

$$\mathrm{d}\sigma = \left(\frac{\partial \varphi}{\partial O}\right)_q \mathrm{d}q\,. \tag{4.30}$$

On the other hand, it is known from electrostatics that the electrical

* Usually in the form

$$\left(\frac{\partial \sigma}{\partial \varphi}\right)_{O,T} = \left(\frac{\partial q}{\partial O}\right)_{\varphi,T}$$

capacity C of a sphere of radius r in a medium with the dielectric constant $\varepsilon = 1$ is

$$C = \frac{q}{\varphi} = r. \tag{4.31}$$

Hence

$$\varphi = \frac{q}{r}; \quad \left(\frac{\partial \varphi}{\partial O}\right)_q = \left(\frac{\partial \varphi}{\partial r}\right)_q \frac{dr}{dO} = -\frac{q}{r^2}\frac{1}{8\pi r};$$

and

$$d\sigma = -\frac{1}{8\pi r^3} q\, dq.$$

After integration and taking into account that $\sigma = \sigma_0$ at $q = 0$ (where σ_0 is the surface tension of an uncharged drop), we obtain

$$\sigma = \sigma_0 - \frac{q^2}{16\pi r^3}. \tag{4.32}$$

It is further necessary to find the vapour pressure of the charged drop. It should be borne in mind that in this case we are interested in the vapour pressure of a drop whose charge q (corresponding to the ion on which the drop has formed) is constant. Taking also into account the fact that the composition of the bulk phases remains unchanged and that the charged component remains at the surface, we can repeat for this case the same considerations as in the derivation of the Thomson–Gibbs equation. The same expression is finally obtained,

$$RT \ln \frac{p_r}{p_\infty} = \frac{2\sigma}{r}\bar{v}, \tag{4.33}$$

but in this case σ is the surface tension of the charged drop [eqn. (4.32)], and therefore

$$p_r = p_\infty \exp\left[\left(\frac{2\sigma_0}{r} - \frac{q^2}{8\pi r^4}\right)\frac{\bar{v}}{RT}\right] \tag{4.34}$$

Thomson obtained this expression by another method. It is represented graphically by the solid line in Fig. 24, in which the dotted-line curve shows the vapour pressure of the uncharged drop [eqn. (4.17)]. It is readily seen that at supersaturation in excess of the supersaturation which corresponds to the maximum value $p = p_m$, each ion will form a drop since in this case there exists no region of values at which the vapour pressure of the drop is higher than that of the ambient medium, and no fluctuations

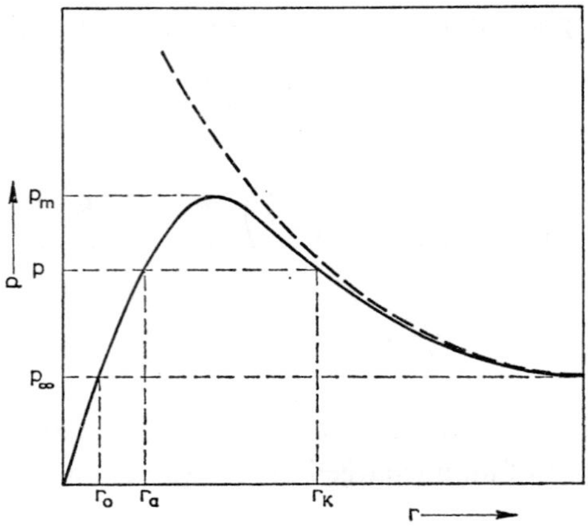

Fig. 24. Dependence of the equilibrium vapour pressure of the drop on its size; —— charged drop, - - - uncharged drop.

are required for the drop to be formed. This is Thomson's explanation of the formation of tracks of ionizing particles in the Wilson cloud chamber.

Tohmfor and Volmer[34] went further, analysing the process of condensation at pressures below p_m, *e.g.* at p (Fig. 24). At these pressures drops will form in the system spontaneously and without hindrance, corresponding to the left-hand limb of the curve, *e.g.* drops of radius r_a. Even at the pressure p_∞, *i.e.* in the absence of supersaturation, drops of radius r_0 will form in the vapour. The system will therefore contain as many drops of radius r_a as there are ions which bring about the possibility of their formation. However, these drops are not yet nuclei, since they are unable to grow further spontaneously and without hindrance. As their radius is increased to r_k, their vapour pressure would exceed that of the ambient medium, as can be seen from the curve under examination. A fluctuation is thus required for the radius to increase from r_a to r_k.

It is obvious from the aforesaid that the formation of nuclei at pressures $p < p_m$ occurs with the aid of fluctuations, and therefore the probability (and consequently also the rate I_1) of the formation of nuclei is given in this case by an equation analogous to (4.22), where the difference ΔF must be calculated for a system with a "sub-nucleus" and with a nucleus. During transition from a system with a drop radius r_a to a system with a drop radius r_k, work $\int_{r_a}^{r_k} \sigma \, \mathrm{d}O$ is performed, where $\sigma = \sigma(r)$, and the free energy also increases due to condensation, by the magnitude $(n_k - n_a)\Delta\mu$, where

$$\Delta\mu = \frac{2\sigma_k}{r_k}\bar{v} = \frac{2\sigma_a}{r_a}\bar{v}\,\dagger\,.$$

Hence the total increase of free energy and the work done by the drop in growing from r_a to r_k will be

$$A_{k_1} = \int_{r_a}^{r_k} \sigma\,\mathrm{d}O - n_k\frac{2\sigma_k}{r_k}\bar{v} + n_a\frac{2\sigma_a}{r_a}\bar{v}.$$

Using Thomson's formula (4.34) for the charged drop and taking into account (4.22) this expression yields

$$A_{k_1} = \tfrac{4}{3}\pi\sigma_0\,(r_k^2 - r_a^2) - \tfrac{2}{3}q^2\left(\frac{1}{r_a} - \frac{1}{r_k}\right) = A_{k_0}\left[1 - \left(\frac{r_a}{r_k}\right)^2 - \right.$$

$$\left. - \frac{q^2}{2\pi r_k^2\sigma_0}\left(\frac{1}{r_a} - \frac{1}{r_k}\right)\right]. \qquad (4.35)$$

This expression, which was obtained by Tohmfor and Volmer by the method of Becker and Döring[26], enables the rate of nucleus formation on ions at $p < p_m$ to be calculated as a function of supersaturation. The ultimate equation, which expresses A_{k_1} in terms of supersaturation, has a very complicated form; we shall therefore be satisfied with the expression (4.35) and write

$$I_1 = k_1 \exp\left(-\frac{A_{k_1}}{kT}\right); \qquad (4.36)$$

since the last two terms in (4.35) are positive, we have $A_{k_1} < A_{k_0}$.

In view of considerations analogous to those expressed in connection with the coefficient k_0 in (4.24), we can affirm that k_1 is proportional to the number of ions which are the centres of condensation.

The process of nucleus formation on ions, so important in the case of the Wilson chamber, apparently does not play a significant role in processes taking place in the atmosphere, since there exist in it condensation centres which ensure the beginning of condensation at a relatively low supersaturation.

The theory of condensation on centres is highly complex, since both

† $\Delta\mu$ is referred to μ_∞ at $p = p_\infty$.

the structure and properties of these centres are complicated and varied. The simplest case of condensation on centres [Krăstanov (1941)[36]] is that of a spherical and completely wettable centre, which can therefore be considered as a "sub-nucleus". In this case the rate of formation of nuclei I_2 is proportional to the number of these sub-nuclei in 1 cm^3. The energy barrier, measured by the work of nucleus formation on the wettable centres is calculated, just as in other already analysed cases, as the sum of the "surface" and "bulk" work values of its isothermal formation:

$$4\pi r_k^2 \sigma - 4\pi r_0^2 \sigma - \frac{2\sigma}{r_k}\left(\tfrac{4}{3}\pi r_k^3 - \tfrac{4}{3}\pi r_0^3\right)$$

where r_0 is the radius of the condensation centre. In this case the surface tension of the wetted centre is considered to be equal to the surface tension of the nucleus. After a few transformations we obtain, taking into account (4.22),

$$A_{k_2} = A_{k_0}\left[1 - 3\left(\frac{r_0}{r_k}\right)^2 + 2\left(\frac{r_0}{r_k}\right)^3\right] \tag{4.37}$$

Since $r_0 < r_k$, $r_0/r_k < 1$ and $(r_0/r_k)^2 > (r_0/r_k)^3$, the expression in square brackets is less than one, and the difference increases with increasing r_0/r_k. This means that A_{k_2} is less than A_{k_0}, the difference being greater the closer r_0 is to r_k. At $r_0 = r_k$ we have $A_{k_2} = 0$, just as one would expect; in this case the nuclei are "ready", and the condensation process occurs without fluctuations. In the opposite case, when the centre is very small ($r_0 \to 0$), we have $A_{k_2} \to A_{k_0}$, *i.e.* the process is virtually not facilitated by the condensation centres.

Further development of the theory of condensation presupposes a more accurate allowance for the properties of condensation centres. Quite recently [Krăstanov (1957)[37], Kaishev and Mutafchiev (1959)[38]] analysed also the case of partial wetting of the sub-nucleus by the material of the new phase to be formed. In this case the activity of the centres depends on the degree of their wettability, *i.e.* on the contact angle.

Numerous problems such as *e.g.* the allowance for the solubility of the centres, the specific properties of the thin adsorption-wetting liquid layers (which coat them partially or completely) the effect of the crystallographic properties of solid centre, and the magnitudes of the charges (if any), have so far remained unresolved.

Extraneous bodies (heterogeneity) can accelerate the process of the

References p. 139–140

formation of the new phase by condensation, and not only by acting as condensation centres in the above sense. As was shown by Volmer[23], the work of nucleus formation is reduced on a completely or partially wettable phase surface. Thus a solid base, which is generally not homogeneous, has a number of areas distinguished by different activities with regard to promotion of the phase-formation process. Such active spots can be at the same time particularly active adsorbents. The presence of small amounts of adsorbed substances can therefore substantially retard the rate of formation of nuclei by blocking the active centres on the solid surface, as demonstrated recently by Mutafchiev and Toshev (1960)[39] in the case of electrolytic deposition of a metal on an extraneous electrode.

Ottewill and Matijevič[40] have published some interesting work on the condensation of water vapour on AgI (and other particles) in cloud seeding, and the influence of adsorbed surface active agents on the crystal form in freshly prepared AgI suspensions.

Another problem related to the formation of nuclei is that of the effect of other components if the initial homogeneous system is a multicomponent one. We have analyzed one very simple case involving an "electrical" component. Much more complicated is the analysis of the case when the second component is distributed over both phases. A detailed study of such a case was carried out by Döring and Neumann in 1940. It would probably be useful to extend this analysis to the condensation on centres, since the latter are either soluble substances (microcrystals of salts) or contain such substances. A systematic and exhaustive analysis of these more complex cases is hindered by the fact that the process of nucleus formation occurs very rapidly, and therefore in the presence of several components their equilibrium distribution cannot be established.

In conclusion, it should be noted that all concepts referred to above neglect the deviation of the properties of small phases from those of the large phase. Since the process of phase formation is determined by the rate of formation of very small particles, it is quite natural that the effect of such deviations should be assumed. In 1958–1962 Shcherbakov *et al.*[41] carried out a thermodynamic analysis of the rate of nucleus formation in a homogeneous phase, on the basis of the above-mentioned correction of the Thomson–Gibbs equation, and showed that if this correction is allowed for, a maximum supersaturation can take place, at which the work of nucleus formation is reduced to zero.

Calculations performed by Nikolov (1959)[42] on the decrease in surface tension for small droplets (*via* an integration of van der Waals' forces) showed that the correction in the Thomson–Gibbs equation, and consequently the influence of long range forces on the phase formation, is quite small.

Further development of these ideas should also involve the formation of nuclei on the substrate and on centres of condensation, where the role of the characteristic properties of small phases might be more accentuated owing to their additional effect in thin layers, which has not so far been taken into account in such phenomena.

3. Liquid Surfaces

Surface tension is measured as energy per unit area or, formally, as the magnitude of force per unit length. The latter is sometimes made use of to draw an incorrect conclusion, *i.e.* that the surface layer is endowed with elasticity and therefore tends to decrease its area. This idea contradicts the molecular kinetic picture of the structure of liquids. There is no reason why molecules in the surface layer should lose their mobility (characteristic of liquids) and the increase of the free surface cannot therefore be visualised as "stretching", *i.e.* as an increase of distance between the molecules in the surface layer. The increase of area takes place as a result of the appearance of new molecules coming up from the bulk of the liquid, indicating that the surface layer is ideally inelastic. This is also borne out by the fact that surface tension of pure liquids does not depend on the magnitude of the area. Indeed, if the surface layer were elastic and surface tension was the result of this elasticity, then during the "stretching" of the surface layer, one should observe a variation of the stretching force (Hooke's law) just as in the case of elastic bodies, and surface tension should depend on the area.

The above ideas can be expressed as follows:

$$\sigma = \frac{\partial A}{\partial O} = n(\varphi^* - \varphi), \tag{4.38}$$

where A is the work of isothermal and reversible increase of the surface, n is the number of molecules which form 1 cm^2 of surface, and $(\varphi^* - \varphi)$ is a measure of the work performed when one molecule is transported from the bulk to the surface.

Another important result is the fact that molecules in the surface

References p. 139–140

layers of pure liquids are involved in as intensive a thermal motion as molecules in the bulk of the liquid. It was on this basis that the theory of "Brownian" waves on the surface of a liquid, mentioned in the preceding section, was developed.

The occurrence of intense motion in the surface layer of liquids can be demonstrated as follows. The number of collisions of gas-phase molecules per second, on a surface of 1 cm² is according to the kinetic theory

$$\frac{p}{\sqrt{2\pi mkT}}.$$

where p is the vapour pressure of the liquid at temperature T, and m is the mass of a molecule.

Since not every colliding molecule is condensed, the number of molecules which pass from the gaseous phase into the surface layer is equal to the above number multiplied by the so-called condensation coefficient α.

At equilibrium the number of condensing molecules is equal to the number of molecules which evaporate, so that the number of molecules which evaporate during 1 sec from 1 cm² of the liquid surface (called the absolute rate of evaporation) is

$$v_{\text{abs}} = \frac{\alpha p}{\sqrt{2\pi mkT}} \tag{4.39}$$

For water at 20° C, $\alpha = 0.03$ and $p = 17.5$ mm Hg; this yields $v_{\text{abs}} = 0.3 \cdot 10^{21}$ molecules per second from 1 cm² of surface. If the size of water molecules is taken into account, we find that 1 cm² can accommodate simultaneously not more than 10^{15} molecules. It follows that a molecule can remain in the surface layer, on the average, for only $3 \cdot 10^{-6}$ sec!

If evaporation takes place in vacuum, the absolute rate of evaporation can *e.g.* be measured from the quantity of vapour evolved under these conditions. Such measurements, carried out by Knudsen (1915)[43] with mercury, gave a good agreement with eqn. (4.39), which later received his name.

4. Adsorption

The phenomenon of adsorption is of particular interest in the physical chemistry of surface and disperse systems. This phenomenon is highly complex from the point of view of the molecular theory which analyses in detail the structure of the adsorption layer. A classical, though rather obsolete exposition of this problem can be found in the previously quoted monograph by Hückel[12].

A particularly objective representation is obtained during the study of liquid surfaces, in which case one of the most complicating characteristics, *i.e.* the specific structure and inhomogeneity of solid surfaces, need not be considered. Moreover, it becomes possible in this case to measure directly the surface tension (solutions) or the two-dimensional surface pressure (insoluble monolayers), which are the most valuable thermodynamic characteristics of the state of the adsorption layer. In the present brief account attention will therefore be confined to the analysis of adsorption on liquid surfaces.

As has already been mentioned in Section 1 [see especially equation (4.18), p. 90], adsorption lowers the surface tension. This lowering is sometimes considerable, and in such cases the adsorbing substance is called surface-active. Obviously, surface-active substances are strongly adsorbing materials. If we confine ourselves to these cases, which are very frequently encountered in practice, we can make use of the simplified ideas of ideal solutions to describe the bulk part of the system in equilibrium. For a molecular interpretation of the phenomenon it is highly important to know whether the adsorption layer is monomolecular or multimolecular. The most important contribution of Langmuir (1916)[44] in this field is the idea that adsorption is in most cases of monomolecular nature. On the basis of this assumption Langmuir developed a highly useful molecular-kinetic theory, the final conclusion of which can be presented in the form of the so-called Langmuir adsorption isotherm

$$N_a = N_s \frac{Kc}{1 + Kc} \tag{4.40}$$

where N_a is the number of molecules adsorbed on 1 cm^2, N_s the maximum possible value of N_a, c the concentration of the adsorbed substance in the bulk phase, and $1/K$ [as may be seen from (4.40)] is equal to the concentration $c_{\frac{1}{2}}$ at which $N_a = \frac{1}{2}N_s$. Eqn. (4.40) was derived for the adsorption from the gas phase on to a solid surface. Let us repeat briefly this elegant calculation in order to estimate the applicability of (4.40) to the adsorption from a solution onto a liquid/gas interface. Langmuir assumed that a certain number N_s of adsorption centres is available in every square centimetre of a solid adsorbent. At equilibrium, a certain fraction θ of these centres is occupied by N_a adsorbed molecules, so that $\theta = N_a/N_s$. The number of molecules desorbed every second from 1 cm^2 will obviously be $\theta N_s \nu^{\dagger} \exp(-\varphi/kT)$, where $\nu \exp(-\varphi/kT)$ is the desorption probability (φ is the activation energy of desorption). According to the molecular-kinetic theory, the number of molecules which fall from the gaseous phase on to one square centimetre is $p/\sqrt{2\pi mkT}$, and only molecules which fall on the unoccupied active surface $(1-\theta)$ will be adsorbed; their number will thus be $\alpha(1-\theta)p/\sqrt{2\pi mkT}$. At equilibrium

$$(1-\theta)\frac{\alpha p}{\sqrt{2\pi mkT}} = N_s \theta \nu \exp(-\varphi/kT),$$

so that the required number of molecules adsorbed per unit surface area is

$$N_a = N_s \frac{p}{p + N_s\sqrt{2\pi mkT}\cdot\frac{\nu}{\alpha}\exp(-\varphi/kT)},$$

since $\theta = N_a/N_s$. Assuming further that $p = ckT$, we obtain (4.40) in which

$$K = \frac{1}{c_{\frac{1}{2}}} = \frac{kT}{N_s\sqrt{2\pi mkT}\cdot\frac{\nu}{\alpha}\exp(-\varphi/kT)}. \tag{4.41}$$

Several experimental adsorption isotherms are very satisfactorily described by the Langmuir equation. In other cases corrections must be applied to account for multimolecular adsorption, lateral interaction, heterogeneity of the adsorption energy etc. These features will not be discussed in this book.

It is obvious that on a liquid surface all places are equally susceptible to adsorption, and therefore N_s corresponds to the state when the entire surface is occupied by the adsorbed molecules. In case of adsorption on active points, especially when their number is not great, the mean distance between the adsorbed molecules remains large, even at $N_a = N_s$. Interactions between adsorbed molecules can therefore be disregarded, and it can be assumed, as before, that the energy of adsorption is independent of the degree of occupation, θ. It is evident that as adsorption on a liquid surface increases, the mean separation of adsorbed molecules can decrease to any extent, until dense packing is attained. It follows that, when

† ν is here a constant of proportionality determining the probability of desorption.

applying Langmuir's isotherm to the adsorption on a liquid surface, we neglect the forces of interaction between neighbouring adsorbed molecules and consider the adsorbed layer to be ideal; however, at the same time we make use of the idea that the limit of adsorption corresponds to the complete covering of the surface, and we therefore take into account the molecular volume which determines N_s. This compromise solution of the problem can yield correct results when the correction for the volume of the molecule in the adsorption layer (allowance for the repulsion forces) is considerably greater than the correction for the forces of attraction. Generally speaking this is not very probable, in view of the greater decrease of the former with increasing distance. The influence of these forces can be taken into account, for example, by choosing a suitable relationship between φ and θ. It will be demonstrated below that such corrections may occasionally yield satisfactory results. Yet another feature remains at present nearly unaccounted for. During adsorption on a liquid surface, the magnitude N_s is determined by the surface area occupied by the adsorbed molecule. If this molecule is non-spherical, as is generally the case with surface-active substances, and has *e.g.* a highly elongated shape, the area it occupies on the surface depends on its position, *i.e.* whether it is horizontal or vertical. The position can, however, vary as the layer is filled up, passing gradually from the horizontal to the vertical, the latter orientation being more economical at greater degrees of occupation. This means that N_s can be an increasing function of θ.

The analysis of adsorption from solution usually boils down to an investigation of the dependence of its surface tension on the concentration of the surface-active component within the liquid. In 1908 Szyszkowski[45] first published the results of such measurements, which he carried out with solutions of the lower fatty acids (C_3 to C_6) in water; his relationship was

$$\Delta = b \ln\left(\frac{c}{a} + 1\right) \tag{4.42}$$

where $\Delta = \sigma_0 - \sigma$ is the lowering of the surface tension of the solution (σ) as compared with the surface tension of water (σ_0) and a and b are constants.

Although Szyszkowski's equation is obeyed in relatively few cases, in view of the typical and relatively simple nature of these cases, it deserves closer attention, and has been theoretically interpreted.

We may begin from the Gibbs adsorption isotherm, which for sufficiently diluted solutions can be represented as (4.18′) [here and in the following text Γ is the excess surface concentration in molecules/cm^2, and Boltzmann's constant k is therefore used in place of R in (4.18′)]

$$\Gamma = -\frac{c}{kT}\frac{\partial \sigma}{\partial c} = \frac{c}{kT}\frac{\partial \Delta}{\partial c}.$$

To obtain the relationship $\Delta = \Delta(c)$ we need another equation, *e.g.* $\Gamma = \Gamma(c)$. For this purpose we can use (with the reservations referred to before) the Langmuir isotherm, by assuming $N_a = \Gamma$. This simplification is admissible particularly for strongly adsorbed components, since when the excess Γ per cm^2 is large, it virtually coincides with the complete surface concentration N_a, and therefore (4.40) can be written as follows

$$\Gamma = \Gamma_\infty \frac{Kc}{1+Kc}, \tag{4.40'}$$

where Γ_∞ corresponds to N_s. By equating both isotherms we obtain

$$\mathrm{d}\Delta = \Gamma_\infty kT \frac{K}{1+Kc}\mathrm{d}c,$$

which yields after integration (and since $\Delta = 0$ at $c = 0$)

$$\Delta = \Gamma_\infty kT \ln(Kc+1) \tag{4.43}$$

or, by comparing it with Szyszkowski's equation,

$$b = \Gamma_\infty kT \quad \text{and} \quad a = \frac{1}{K}. \tag{4.44}$$

We thus have a probable interpretation of the experimental constants b and a, in good qualitative agreement with experiment. The experimentally determined values of b for various homologous compounds are close to each other, which is quite natural since $b = \Gamma_\infty kT$ is determined by the area $1/\Gamma_\infty$ per molecule of the adsorbed substance. The values $1/\Gamma_\infty$, calculated as kT/b, amount usually to 20–40 Å^2 for various molecules, which is in good agreement with other data relating to these dimensions, and corresponds to a certain, most frequently "vertical" orientation of molecules in the adsorption layer. This is also probable since these molecules have usually a highly asymmetric structure and one of their ends is very strongly lyophilic (*e.g.* —COOH or —OH). Numerous studies of the orientation and structure in the adsorption layer of surface-active agents have been carried out on this basis.

More reliable values of Γ_∞, or of the area per molecule $1/\Gamma_\infty$, can be obtained from determinations of $\Delta = \Delta(c)$, using only the Gibbs isotherm, according to which $\Gamma kT = \partial\Delta/\partial \ln c$ is the gradient of the curve $\Delta = \Delta(\ln c)$. If we make use of $\Delta = \Delta(c)$ up to values for which $\partial\Delta/\partial \ln c$ attains a maximum and becomes independent of c, or if we extrapolate reliably to this zone, we can obtain Γ_∞ directly, without resorting to supplementary considerations such as Langmuir's isotherm. Since in this case the concentrations become fairly high, it is frequently necessary to replace c by activities. A large volume of critically treated material with respect to $1/\Gamma_\infty$ has been collected by Taubman (1950)[46] for the aqueous solutions of the medium- and low-molecular surface-active substances. On the strength of these data Taubman proposed the following generalization for the magnitude of $1/\Gamma_\infty$ in square angstroms.

	Acids	*Alcohols*	*Amines*
Aliphatic	30.5	28.5	30.0
Aromatic	30.0	28.5	28.0

From the standpoint of the developed interpretation of Szyszkowski's equation it also becomes obvious why even the neighbouring members of the same homologous series have widely differing values of a when the values of b are very close. According to (4.44) and (4.41), $a \sim \exp(-\varphi/kT)$, so that even small differences in φ give rise to widely differing values of a.

An explanation of the familiar Duclos–Traube rule (1878)[47], which had been established experimentally for solutions of the lower homologues of the fatty acids, can also be obtained with the aid of the above-mentioned Langmuir's treatment. According to this rule, the concentration ratio of two neighbouring homologues is constant and equal to approximately 3.2 for the same Δ. According to (4.42), this can be written as follows

$$b_n \ln\left(\frac{c_n}{a_n} + 1\right) = b_{n+1} \ln\left(\frac{c_{n+1}}{a_{n+1}} + 1\right),$$

where n is the number of the homologue. Since $b_n \approx b_{n+1}$, this means that

$$\frac{a_n}{a_{n+1}} = \frac{c_n}{c_{n+1}} = 3.2$$

References p. 139–140

according to the Duclos–Traube rule. If we disregard the unimportant difference in the pre-exponential factor in eqn. (4.41), for the n-th and $(n+1)$-th homologue, we obtain from (4.44) and (4.41)

$$\frac{a_n}{a_{n+1}} = \exp\left[(\varphi_{n+1} - \varphi_n)/kT\right] = 3.2 = \text{const.},$$

i.e. the constancy of c_n/c_{n+1} is associated with the constancy of $(\varphi_{n+1} - \varphi_n)$. Such constancy in the increase of the adsorption potential from one homologue to another could be explained by assuming that each fatty acid molecule lies in the surface, so that its bond with the "surface" increases by the same amount when the molecule is lengthened by one CH_2 group; according to this argument $(\varphi_{n+1} - \varphi_n)$ would be the fraction of the adsorption potential corresponding to one such methylene group.

This explanation is, however, in opposition to the fact that values of Γ_∞ for the same objects correspond not to the "lying" (horizontal), but to the "erect" (vertical) position, and are thus almost independent of n. This contradiction may be resolved by assuming that for small covering ratios of the adsorption layer, when the Duclos–Traube rule is obeyed, the adsorbed molecules lie in the surface, whereas with increasing layer thickness the molecules gradually pass into the erect position. It should, however, be noted that this treatment is strictly incompatible with the use of Langmuir's isotherm, in which it is assumed that Γ_∞ is constant and independent of the extent to which the adsorption layer is filled.

The next step in the development of the problem consists of taking into account the forces of attraction between adsorbed molecules or, briefly, the energy of interaction between the adsorbed substances. This was done by Frumkin in 1925[48], for the region under consideration.

Expression (4.43) can also be written in the form $\Delta = \Delta(\Gamma)$, correspondingly to (4.40′), *i.e.*

$$\Delta = -\Gamma_\infty kT \ln\left(1 - \frac{\Gamma}{\Gamma_\infty}\right). \tag{4.45}$$

This equation is the conventional equation of state of the adsorption layer, which relates $\Delta = \pi$ (where π is the two-dimensional pressure of the adsorption layer as will be shown in the analysis of monomolecular layers of insoluble substances on a liquid surface) to the surface concentration or to the surface molecular "volume"—the molecular area $1/\Gamma$. Since we have now taken into account the area occupied by the molecules, this equation is analogous to the equation of state of a real gas or solution, the attraction forces being, however, disregarded. Frumkin introduced such a term, $-a'(\Gamma/\Gamma_{\infty})^2$, analogous to the corresponding correction in van der Waals' equation, so that

$$\Delta = -\Gamma_{\infty}kT\ln\left(1-\frac{\Gamma}{\Gamma_{\infty}}\right) - a'\left(\frac{\Gamma}{\Gamma_{\infty}}\right)^2. \qquad (4.46)$$

Frumkin's substantiation of (4.46) is more complete. By combining this expression with Gibbs' isotherm (4.18′) we obtain

$$\mathrm{d}\ln c = \frac{\mathrm{d}\Delta}{\Gamma kT} = \frac{\mathrm{d}(\Gamma/\Gamma_{\infty})}{(1-\Gamma/\Gamma_{\infty})\Gamma/\Gamma_{\infty}} - 2\frac{a'}{\Gamma_{\infty}kT}\mathrm{d}(\Gamma/\Gamma_{\infty})$$

or, after integration, assuming $\Gamma/\Gamma_{\infty} = \frac{1}{2}$ and $c = c_{\frac{1}{2}} = 1/K$,

$$\ln Kc = \ln\frac{\Gamma/\Gamma_{\infty}}{1-\Gamma/\Gamma_{\infty}} - \frac{a'}{\Gamma_{\infty}kT}(2\Gamma/\Gamma_{\infty} - 1). \qquad (4.47)$$

The required relationship $\Delta = \Delta(c)$ is thus obtained in the parametric form (4.46) and (4.47). It should be stressed here that eqn. (4.46), just like the three-dimensional van der Waals equation of state, comprises also the unstable heterogeneous zone (occurring initially by the formation of dimers and bigger aggregates at higher concentrations) which corresponds

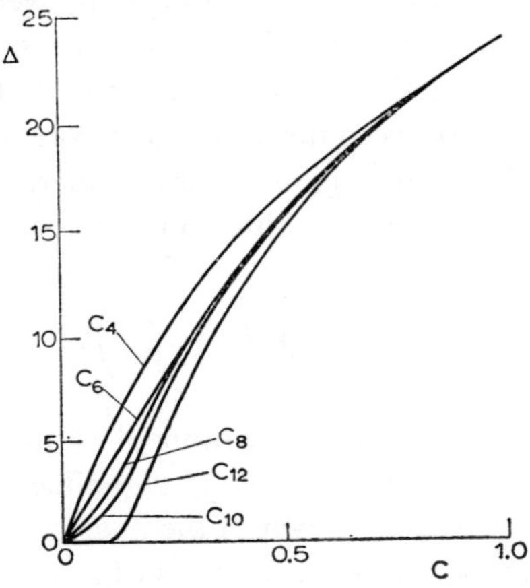

Fig. 25. Dependence of the lowering of surface tension on concentration, for aqueous solutions of fatty acids.

References p. 139–140

to the formation of the dense (condensed) adsorption layer. The actual shape of the $\Delta = \Delta(\Gamma)$ curve, and consequently of the $\Delta = \Delta(c)$ curve, is found conventionally by drawing for the first of the two relationships a straight line in the heterogeneous zone [zone of the maximum and minimum of curve (4.46)] in such a way as to make the area bounded by the curve above it equal to the corresponding area below it. As a result of such operations we obtain ultimately the curves $\Delta = \Delta(c)$ shown in Fig. 25 (Frumkin interpreted the experimental data of Szyszkowski (C_4 and C_6) and his own (C_8, C_{10} and C_{12}) for the corresponding fatty acids), in which C_4 corresponds to Szyszkowski's equation and C_{10} has a clearly expressed concavity characteristic when attractions between adsorbed molecules are considered, which passes in C_{12} into the extreme case when measurable Δ values occur only at considerable values of c. For the sake of comparison, the curves for the various substances are arranged so that $c = 1$ for the same Δ (= 24 dynes/cm). The corresponding concentrations are: lauric acid, 5.9×10^{-6}; capric acid, 9.5×10^{-5}; caprylic acid, 11.2×10^{-4}; caproic acid, 21.2×10^{-3}; butyric acid, 26.7×10^{-2} (all in moles/litre).

Qualitative interpretation of eqns. (4.46) and (4.47) presents certain difficulties; on the one hand the attraction forces increase the adsorption by drawing the adsorbed molecules into the adsorption layer, owing to the energy of the adsorbate–adsorbate interactions, thus contributing to the increase of Δ, and on the other hand the same attraction reduces the two-dimensional pressure, *i.e.*, Δ. It follows from the thermodynamic considerations that the latter effect is always predominant, so that any correction for the existence of attractive forces between adsorbed molecules should result in the lowering of Δ and in the observed concavity of the adsorption isotherm, and correspondingly of the isotherm $\Delta = \Delta(c)$. The curves in Fig. 25, obtained experimentally by Frumkin for the indicated higher fatty acids, corroborate the theory very closely, and enable the adsorbate–adsorbate interaction energy expressed by the constant a' in (4.46) and (4.47) to be estimated*.

This does not however, even qualitatively exhaust the diversity of experimental results relating to the function $\Delta = \Delta(c)$. For the higher-

* A number of authors [Volmer and Mahnert (1925)[49], Cassel and Formstecher (1932)[50] and others] allowed for the effect of the adsorbate–adsorbate interactions in a form even closer to the van der Waals' equation. All these attempts, as rightly noticed by Adam[2], are subject to an inaccuracy characteristic of any equations of state for real systems.

molecular and complex-structure surface-active substances the isotherms $\Delta = \Delta(c)$ present a more complicated form. This complexity is manifested in the first place by the so-called micelle-formation at higher concentrations, *i.e.* by the formation of more complex and coarser particles within the liquid. Therefore the concentration of the molecularly dispersed portion of the surface-active substance ceases to increase when its total bulk concentration is increased, since the added excess is used to form micelles; constancy and independence of the surface concentration and of Δ on the total bulk concentration are thus attained. In view of the irreversibility or incomplete reversibility of the process of micelle-formation, these bulk effects can result in supersaturation phenomena with respect to the molecularly dispersed portion, which begin to decrease when a sufficient degree of supersaturation has been attained, as a result of the increased rate of formation of micelles. This may even cause an inversion of the relation between Δ and c.

Ekserova and the present author have recently (1963)[51] succeeded in revealing a highly interesting variation of the function $\Delta = \Delta(c)$ for many surface-active agents of the foaming agent type, present in very low bulk concentrations (in a number of cases one or more orders of magnitude lower than the micelle-formation concentration). These peculiarities could not be accounted for by the bulk effects described above. Fig. 26 shows such a typical curve for OP20 (octylphenol polyglycol ether, $n = 20$) for $c < 2.5 \cdot 10^{-5}$ moles/l, the micelle-formation concentration being in this case $\approx 4 \cdot 10^{-4}$ moles/l.

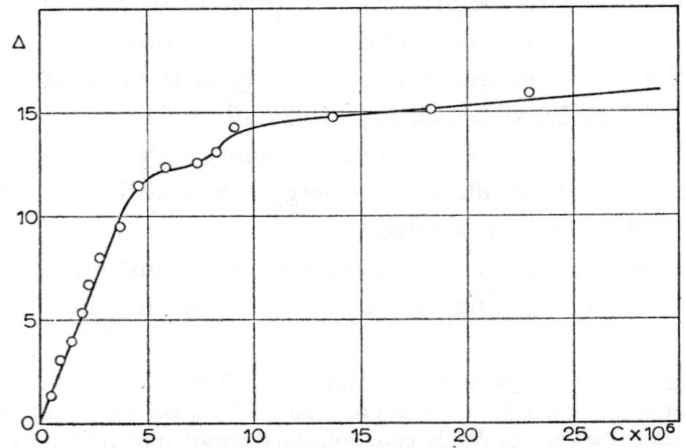

Fig. 26. Dependence of the lowering of surface tension on concentration, for aqueous solutions of OP20.

The peculiarity of these curves consists in the fact that at the beginning $\Delta \sim c$; for the lower homologues, the curve is even convex, this being followed by a double discontinuity which is expressed more clearly in the higher homologues. Finally, the curve approaches gradually and very gently to a constant value of Δ which corresponds to the concentration at which micelles are formed. It seems as if the curve in Fig. 26 consists of two sections. Indirect data, which will be discussed in the chapter dealing with foams and emulsions, suggest that the dense monomolecular adsorption layers form in the discontinuous regions of these curves, so that at greater values of c only multimolecular adsorption or rather reorganisation of the monolayer can occur.

5. Determination of Surface Tension of Liquids

The tendency to characterise the surface properties of pure liquids and solutions of surface-active substances by their surface tension is explained by the fact that the latter property can be measured relatively easily and with fairly high accuracy. The numerous available methods of measuring surface tension can be divided into static and dynamic. In the static methods the measurement is carried out with a stationary liquid, the surface being formed before the measurement. Under these conditions it can be assumed that equilibrium has been established between the solutes (in the case of solutions) in the bulk and on the surface of the liquid. The capillary rise method is a typical example of this class of surface tension measurements.

In case of the dynamic methods, in which the surfaces are formed immediately prior to measurement, adsorption equilibrium cannot always be established*. Therefore the dynamic methods, which in most cases require less time and are more convenient than the static ones, do not always yield the equilibrium value of surface tension. Measurements of the dynamic, non-equilibrium surface tensions are sometimes of special interest, particularly in the study of adsorption kinetics. A brief analysis will now be given of some of the methods used most frequently. More detailed information on the present state of this problem can be found *e.g.* in refs. 2 and 3.

* In the adsorption layers of medium- and high-molecular surface-active substances there also occurs a slow rearrangement of the layer structure, sometimes referred to as "ageing"; this is also associated with a change of surface tension with time.

The capillary rise method

This is one of the most frequently used and most accurate methods of measuring the surface tension of liquids. Its theoretical basis is highly reliable and has been thoroughly developed. The measurement is based on eqn. (4.16′) for the capillary pressure.

The boundary between the liquid and gaseous phases is formed by plane surface A″A and A′A‴ and by the concave surface BB′ in the capillary tube PP′QQ′ having an inside radius r (Fig. 27). If we consider the surface BB′ to be a part of a sphere whose centre is O, then the angle θ between the direction OX (parallel to the horizontal planes A″A and A′A‴) and OB is equal to the contact angle θ between phase 1 and the vertical wall of the tube at the point of contact, B.

In this case the radius of curvature of BB′ is $r/\cos\theta$. If the pressure above the plane surfaces AA″ and A′A‴ is p_0, it is clear that the pressure in the capillary tube at the level AA′ is also p_0. Then the pressure p_1 in the liquid 1 at level BB′ will be

$$p_1 = p_0 - \rho_1 gh$$

and in the gaseous phase 2, at the same level,

$$p_2 = p_0 - \rho_2 gh,$$

where ρ_1 and ρ_2 are respectively the densities of the liquid and of the gas, g is the acceleration due to gravity, and h is the height of the liquid column in the capillary tube. According to (4.16′)

$$p_2 - p_1 = \frac{2\sigma}{r}\cos\theta$$

which yields

$$\frac{2\sigma}{r}\cos\theta = (\rho_1 - \rho_2)gh. \tag{4.48}$$

If we know r, θ, ρ_1 and ρ_2, and if we measure h we can calculate σ.

The capillary rise method is generally employed in the frequent case when the liquid wets the walls of the glass cylinder completely, *i.e.* for $\cos\theta = 1$; eqn. (4.48) then becomes

$$\frac{2\sigma}{\rho_1 g} = rh = a^2, \tag{4.49}$$

neglecting ρ_2 which is small compared with ρ_1.

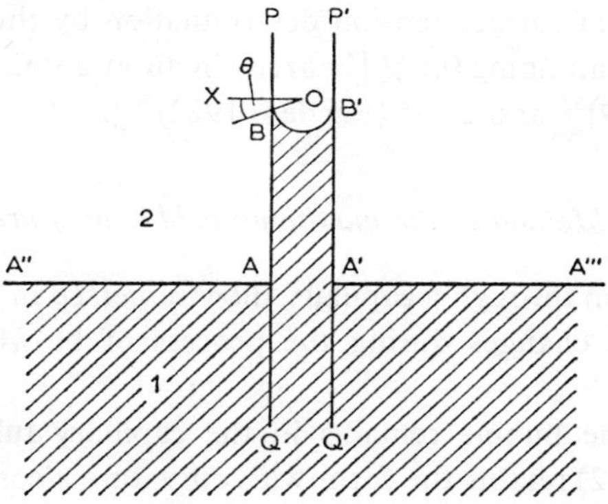

Fig. 27. Surface tension measurement (capillary rise method).

In a more detailed analysis, eqns. (4.48) and (4.49) are found to be inaccurate, since the weight of the meniscus has not been allowed for, (h being the height from the level of plane liquid to the lowest point of the meniscus) and because the gravity-induced deformation of the meniscus has been neglected. It is easy to see that both these errors increase with increasing radius of the capillary. The first correction was derived by Rayleigh (1915)[52] in the form (for $\theta = 0$)

$$a^2 = r(h + r/3).$$

The correction for meniscus deformation was found to be more complicated. It can be represented either in the form of an approximate equation (Rayleigh) or by another approximation, as carried out by Bashforth and Adam[53] and finally by Sugden (1921)[54], in the form of correction tables in which the correction is calculated for various values of r/a.

No experimental details of this and of the other methods can be included in the present brief account. The diverse versions of the capillary rise method can be reduced to finding conditions at which h and r can be measured as accurately as possible. For instance, when determining h (usually with a cathetometer) it is difficult to fix accurately by an optical method the level of a liquid in a wide vessel, and the difference of levels between a narrow and a wide capillary tube is therefore measured, which is related correspondingly to the difference of capillary pressures. The problem of the exact determination of r is also a subject of detailed discussion, in view of the non-uniformity of the cross-section area along the capillary, and of the ellipticity of this section.

The accuracy of surface tension determination by this method is high for pure liquids, attaining 0.1 % [benzene in air at 20° C is 28.88 (Harkins and Brown, 1919)[55] and 28.85 (Sugden, 1922)[56]].

Method of the maximum bubble pressure

When air is blown through a capillary tube immersed in a liquid, the form of the meniscus changes during the process of blowing, as shown in Fig. 28.

As long as the bubble remains in the capillary tube, the meniscus (positions 1 and 2) retains the form with curvature $r/\cos\theta$. The curvature then begins to increase until hemispherical form (3) is reached ($\cos\theta = 1$), after which it again begins to decrease (4 and 5). The external pressure will therefore be equal to the hydrostatic pressure, proportional to the lowering of the level, plus the capillary pressure which is constant at ($2\sigma/r \cos\theta$), as long as the lower end of the capillary tube has not been reached; it then begins to increase to the value $2\sigma/r$, and finally decreases again. It is clear that when the pressure corresponding to the hemispherical meniscus has been reached, subsequent inflation of the bubble occurs spontaneously until it breaks away. The maximum pressure in the process of bubble blowing is therefore determined by the hydrostatic pressure, which is given by the immersion depth of the capillary tube, and by the capillary pressure $2\sigma/r$ which does not depend on the contact

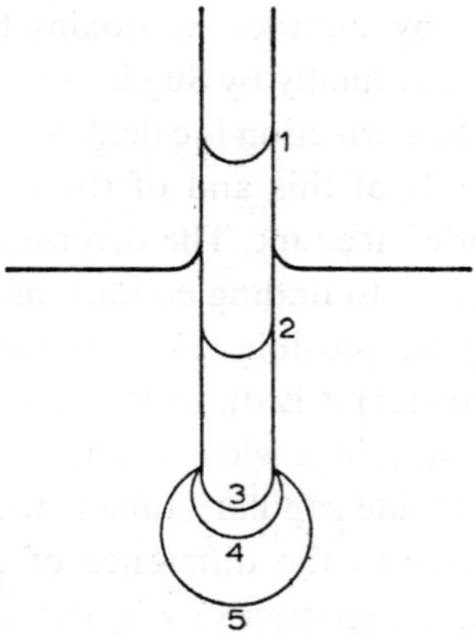

Fig. 28. Successive positions of the meniscus in the capillary tube during bubble blowing in a liquid.

References p. 139–140

angle. The independence of this method of the contact angle is its characteristic feature as compared with the capillary rise method. Hence the maximum pressure attainable when the bubble is slowly inflated will be

$$p = gh\rho_1 + \frac{2\sigma}{r} \tag{4.50}$$

where h is the depth of immersion of the capillary tube in the liquid.

Eqn. (4.50) is only valid for sufficiently slow blowing, since it does not allow for the dynamic resistance during the movement of the liquid, which, just as the corresponding additional pressure which is required to overcome it, tends to zero when the rate of blowing is extremely slow. When applying this method with great accuracy, just as in the determination of σ by the capillary rise method, one should allow for the deformation of the bubble due to the weight of the liquid, and for the weight of the liquid in the meniscus. The corresponding corrections can be found in special literature (Sugden, 1922)[56]. It should be noted that in this case the working surface is freshly formed, and high results might therefore sometimes be obtained in solutions of surface-active substances. The fundamental advantage of this on the whole not very accurate method, *i.e.* its independence of the contact angle, is not in general of great significance, as in most cases conditions of complete wetting can be ensured. The method has nevertheless found wide application in laboratory work owing to purely practical advantages.

In 1921 Sugden proposed an improvement to this method by using two capillary tubes with different radii r_1 and r_2, immersed in the liquid to the same depth, thus making it unnecessary to measure the depth of immersion. If the maximum bubble pressures for the two capillaries are p' and p'' at the same depth, then according to (4.50):

$$p' - p'' = 2\sigma\left(\frac{1}{r_1} - \frac{1}{r_2}\right).$$

A clever modification of this method was suggested by Warren in 1927[57], for the measurement of slight differences in the surface tension of solutions of surface-active substances. Two identical capillary tubes connected to the same suction apparatus are immersed in two vessels containing liquids for which the surface tension difference is to be determined. The depth of immersion in one vessel is adjusted until bubbles begin to escape simultaneously from both capillary tubes, and the surface tension difference is then determined from the difference in the levels of immersion.

The independence of the contact angle renders this method convenient for measuring the surface tension at the boundary of two liquids, since in this case the contact angle of the wall of the capillary tube is not equal to zero and may assume various values. In this case a drop of one liquid is squeezed out into the other liquid (Rebinder, 1926)[58].

The stalagmometric method

Another simple, and consequently widely applied, method of measuring surface tension is the very old stalagmometric method (Tate, 1864)[59], in which surface tension is calculated from the weight of the liquid drop discharged from the capillary tube. The theory of this method has not yet been developed, and the frequently used formula $Q = 2\pi r\sigma$ (where Q is the weight of the drop, and r is the radius of the capillary tube) is highly inaccurate. This equation is derived from the assumption that the drop breaks off when its weight is equal to, or slightly exceeds, the force (due to surface tension) which acts along the perimeter of the hemispherical drop and tends to pull the drop back into the capillary tube. The process of drop break-off is in reality much more complex. Fig. 29 shows diagrammatically a series of photographs made during breaking away of a drop. The unstable state corresponds to the complex drop form with a "waist". Two drops form on breaking away, a large and a very small one, and a convex meniscus remains at the end of the tube. As a result of their experimental investigations, Harkins and Brown (1919)[55] derived the following formula for the weight of a drop which breaks away.

$$Q = 2\pi r\sigma f\left(\frac{r}{v^{\frac{1}{3}}}\right) \tag{4.51}$$

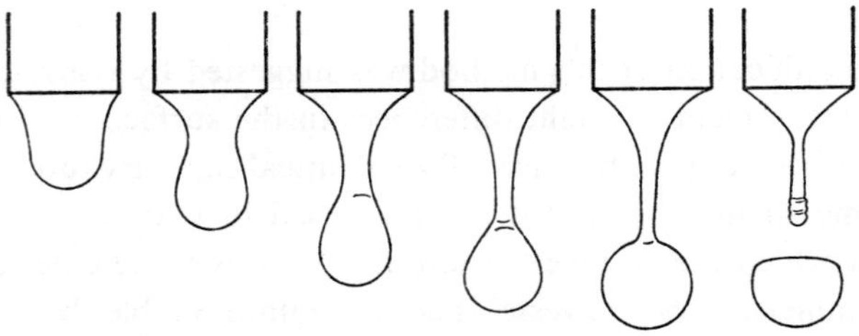

Fig. 29. High speed photographs of a falling drop.

References p. 139–140

where v is the volume of the drop. The values of $f(r/v^{\frac{1}{3}})$ are listed in special tables drawn up by these authors. Use of the formula $Q = 2\pi r\sigma$ may yield values of σ which differ by 10 % or more from the actual ones. The stalagmometer method is sometimes also used for the determination of the liquid–liquid interface tension, in which case the drops of one liquid are allowed to fall in the other liquid.

The "detachment" method

To detach a body from the surface of a liquid which wets the body, it is necessary to overcome the same surface tension forces which operate when a drop is broken away. The methods of determining the surface tension by measuring the force required to detach a body from a liquid are therefore similar to the stalagmometer method described above. However, their advantage over the latter method consists in that they make it possible to choose the most convenient form and size of the body (platinum rod, ring or plate) so as to enable the measurement to be carried out rapidly, but without any detriment to its accuracy. The "detachment" method has found an application in the case of liquids whose surface tension changes with time. Such substances are *e.g.* soap solutions, which had been investigated in detail by Lottermoser (1935)[60] and du Noüy (1926)[61] by means of this method. Tables for the accurate calculation of surface tension by the ring detachment method were proposed by Harkins and Jordan in 1930[62].

Determination from the shape of drop or bubble

A droplet of a liquid or a gas bubble in a liquid are partly deformed from true spherical form by gravitational forces. The degree of deformation is the greater the lesser the surface tension which favours the spherical shape. The gravitational effect which causes the deformation increases with increasing difference in the densities of the drop (or bubble) and the surrounding medium and with the increasing size of the drop (or bubble). Therefore the equilibrium form, which is determined by the condition of the free energy minimum, is a function of surface tension and can be used for its measurement. Methods based on this concept are strictly static and have been widely adopted in particular for the measurements of surface tension of surface-active substances as a function of time

(Nutting and Long, 1941)[63] in spite of considerable experimental difficulties. Since the drop or bubble rest on a solid surface, their equilibrium form is related to the contact angle which is not susceptible to direct and accurate measurement. Recently Smolders (1961)[64] has successfully used the method of the drop and bubble form for a detailed investigation of wetting phenomena.

Dynamic methods

Of the reviewed methods of surface tension determination only the methods of capillary rise of liquids and of the equilibrium form of the drop or bubble are completely static. In all other methods the measurement is associated with a more or less rapid change of surface areas. It is, however, customary to denote as dynamic only those methods in which the surface tension is measured during rhythmic oscillations of the liquid surface. Such rhythmic oscillations appear in jets, during the deformation of drops, and on the surface of a fluctuating liquid. In all these cases the tendency of the liquid to reduce its surface area, measured by the surface tension, counteracts the increase of the surface. If deformation and an increase of surface occur as a result of external forces, the surface tension restores the liquid to its initial state. This state may, however, be overshot as a result of inertia, leading to rhythmic oscillations the investigation of which can yield the magnitude of surface tension. These phenomena and their relationship to surface tension were studied in detail by Rayleigh (1879) and form the basis of the dynamic methods of surface tension measurements.

When the surface of solutions of surface-active substances undergoes deformation, a redistribution of concentration occurs in the adsorption layer, and the surface tension varies from one surface section to another. The local contractions and expansions of the adsorption layer are determined by the mobility of the adsorbed molecules (diffusion along the surface), the mobility of these molecules in the bulk (bulk diffusion), the rate of rise from the bulk up to the surface, and the linking of the adsorbed molecules in the adsorption layer. These interesting phenomena, which had not been allowed for in Rayleigh's theory, are analysed in detail in Levich's "Physico-chemical hydrodynamics"[65].

References p. 139–140

6. Monomolecular Layers of Insoluble Substances on Liquids

The monomolecular layers obtained when very small quantities of insoluble compounds are applied to the surface of a liquid, can be analysed as the extreme case of the monomolecular adsorption layers of solutions, when the solubility of the surface-active substance is so small as to be negligible. Unfortunately, the utilisation of theoretical results obtained for solutions by means of the boundary transition of this kind is rendered difficult by the fact that the theory is based on the existence of a dynamic equilibrium between the bulk and the surface of the liquid, the bulk concentration playing a substantial role; if this concentration is disregarded in the case of insoluble substances, the theory is stripped of this physical basis. Monomolecular layers of insoluble substances are therefore considered separately as an isolated two-dimensional solution.

The field was opened up in 1891 by Pockels[66] who started her investigations by varying the thickness of a very thin layer of oil on water by moving a paper barrier resting on the surface. As will be seen, the idea of this two-dimensional piston proved very fruitful. The oil spot on the water surface could be compressed or expanded. Measurements of the lowering of surface tension as a function of area occupied by the oil spot showed that when the oil film is compressed the surface tension becomes equal to that of water, but experiences a sudden, sharp drop at a certain degree of compression. This observation was related to the phenomenon of quelling stormy seas with oil, which had long been known to sailors. It was found that such "damping" of waves begins when the density of the surface layer assumes the value at which the drop of surface tension is observed.

In 1899 Rayleigh[67] confirmed Pockels' results and supplied an explanation of these phenomena, which was not very far removed from truth. According to him we are dealing with a monomolecular layer of oil. It was found that, if the area occupied by the oil molecules is sufficiently large, these molecules do not interact with each other and do not affect the surface tension. When the area is decreased they begin to repel each other, and finally a force is generated which counteracts the surface tensions, causing the sudden drop. Rayleigh believed that the beginning of the lowering of surface tension corresponded to the dense monomolecular oil layer. As the total area of the film and the amount of oil applied were known it was easy, by assuming that the density of the dense monomolecular layer is the same as the density of oil in bulk, to calculate that the diameter of one molecule of oil (again assuming spherical form) was approximately 10^{-7} cm. The order of the molecular diameter thus obtained is quite reasonable.

These calculations did not allow for molecular thermal motion. This motion underlies the tendency of the film to expand, which should result in a gradual lowering of surface tension long before dense packing is achieved. Since this gradual lowering of surface tension which precedes its sharp drop when dense packing is attained, is rather insignificant in oils, it was not observed by Pockels. The above considerations yield a very simple explanation of the lowering of surface tension, which can have a great theoretical and practical significance.

Surface tension and surface pressure

Rayleigh's elementary calculation showed that is is possible to carry out direct experiments with monomolecular layers supported on the surface of liquids. If the layer is assumed to be homogeneous, this conclusion does not arouse any doubts, and allows a number of conclusions to be drawn with respect to the properties of such a layer. In the first place, in accordance with molecular kinetic ideas, the molecules in the monolayer should move similarly to molecules in the bulk of the solution. We should thus observe a kind of osmotic two-dimensional pressure in the surface layer, which tends to distribute the applied molecules uniformly over the entire surface of the liquid. This two-dimensional osmotic pressure is obviously caused by the random thermal motion of molecules in the monolayer.

If we partition the liquid surface by means of a plate which touches the surface but is not wetted by the liquid (*e.g.* a strip of paper impregnated with paraffin wax following the example of Pockels) (Fig. 30) and if we

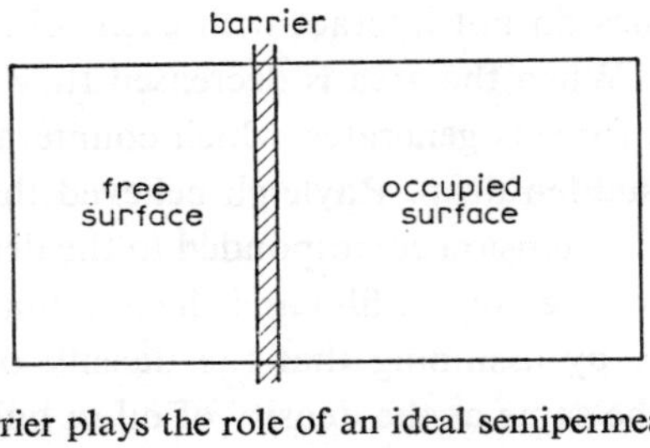

Fig. 30. The barrier plays the role of an ideal semipermeable membrane.

References p. 139–140

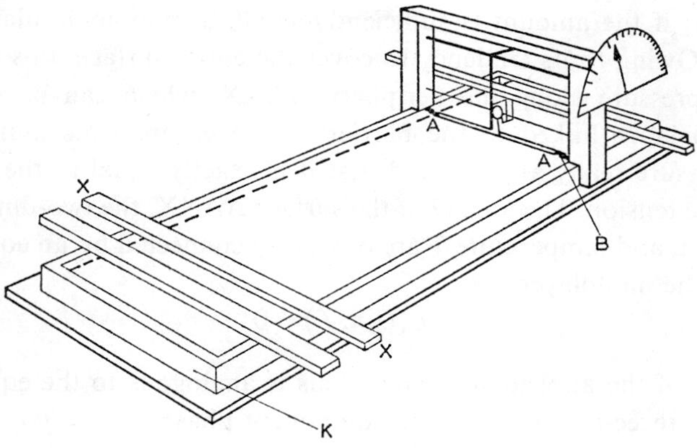

Fig. 31. The Langmuir–Adam Balance.

apply a certain quantity of an insoluble substance to the surface on one side of the barrier, this barrier will play the role of an ideal semipermeable membrane. The insoluble substance cannot pass through the barrier, whereas the molecules of the liquid base are able to pass from one surface to the other by "diving" under the barrier. The two-dimensional osmotic pressure under consideration (*i.e.* the surface pressure π) will therefore exert a pressure on the barrier with a force $f = \pi l$, where l is the length of the barrier. If the barrier is easily movable and linked to a dynamometer or a balance to compensate f, then such an apparatus will enable this force and the corresponding surface pressure to be measured directly. This is precisely the principle of an apparatus designed by Langmuir[68], which was later perfected by many workers, chiefly by Adam[69] (Fig. 31).

A liquid on which the monolayer will subsequently be produced is poured up to the very brim of the trough K. The barrier AA, impregnated with paraffin wax, is fixed to the trough walls by very light, non-wettable and non-sinking threads B, and forms a continuous semipermeable barrier on the surface. This barrier is linked to a sensitive microbalance which enables the unilateral pressure exerted on the barrier to be compensated. Another barrier is used to control the magnitude of area AAXX covered with the monolayer. A certain volume of the substance, dissolved in a volatile solvent but insoluble in the trough liquid is placed in AAXX. This method permits the use of extremely small amounts of the layer substance. When the volatile solvent escapes by evaporation, there remains on the surface AAXX a layer whose quantity is known with precision; if the amount is sufficiently small, a monomolecular layer is formed. Owing to its tendency to cover the entire surface, this layer will exert a pressure along the periphery AAXX, which can be measured by the balance linked to the barrier AA. Measurements indicate that such pressure really appears and that it is exactly equal to the lowering of surface tension. The area O of the surface AAXX, the two-dimensional pressure π, and temperature T are obviously connected by an equation of state of the monolayer, *i.e.*

$$\varphi(O,\pi,T)=0$$

per mole of the applied substance. This is analogous to the equation of state of a three-dimensional one-component phase

$$\varphi(p,v,T)=0$$

per mole. The dependence of the surface pressure $\pi=\pi(O)$ on the area O occupied by the monolayer containing one mole of the substance, or better, on the area which corresponds on the average to one molecule of the monolayer O/MN_{A} at constant temperature, is the isotherm of state of this monomolecular layer.

Before we pass to the description of some results of these measurements, let us consider once more why the lowering of surface tension measured by Pockels is identical with the surface pressure π. Surface tension is the work of the isothermal reversible formation of 1 cm^2 of surface. If the barrier on the surface of the liquid is moved reversibly (very slowly) by a distance $\mathrm{d}x$ (Fig. 30) in the direction of the free part of surface, then the work done by the surface pressure π is $\pi l\,\mathrm{d}x$ (where l is the length of the barrier). On the other hand, if σ_0 denotes the surface tension of the pure liquid, and σ that of the liquid covered with the monolayer, then the area $l\,\mathrm{d}x$ with surface tension σ will be replaced after this movement of the barrier by the same area with surface tension σ_0, and therefore the change of the surface energy will be $(\sigma_0-\sigma)l\,\mathrm{d}x$. By equating $\pi l\,\mathrm{d}x$ and $(\sigma_0-\sigma)l\,\mathrm{d}x$, we obtain $\pi=\sigma_0-\sigma$.

The main advantage of measuring π rather than σ consists in that the sensitivity of π measurements is greater than that of the measurements of surface tension. It has already been indicated that the low sensitivity of σ measurements was the cause of the erroneous conclusion that surface pressure does not appear until the oil monolayer has attained the state of dense packing. The methods of surface tension measurements give at

References p. 139–140

best an accuracy of 0.1 %, which *e.g.* for water ($\sigma_0 = 72$, or of the order of 100 dynes/cm) is about 0.1 dynes/cm. The Langmuir balance has an accuracy which is 10 times as great (0.01 dynes/cm) or even greater for certain (differential) methods. It will be seen that this degree of accuracy is of particular importance for the measurement of the molecular weight of such substances as proteins. This method unfortunately cannot be applied to the easily soluble surface-active substances, owing to their diffusion in the solution beneath the barrier. Finally it should be noted that highly reliable surface pressure measurements can be achieved only if proper precautions are taken against evaporation and contamination of the monolayer with dust particles from the atmosphere.

Investigations of monomolecular surface layers of insoluble and sparingly soluble substances by measuring their surface pressure

These investigations cover a self-consistent field of the physical chemistry of surface, and they have retained their significance up to the present time. This method is currently used for the study of high-molecular substances of the protein type—such monolayers are in fact a simple model for the study of biological membranes. Although the results achieved in this field will not be quoted, some simpler cases will be considered in order to indicate the characteristics of the trend and the significance of these investigations. The reader can find detailed information in the publications already quoted[2,3,70].

It is known that the equation of state combines in the simplest case ("ideal" gases and solutions) the pressure, molar volume and absolute temperature into a equation which is universal for equimolecular quantities. In the case of higher concentrations and at lower temperatures this equation loses its universal character, and the effects of constants specific to each substance, which express the energy of molecular interaction* begin to appear. Investigation of the equation of state of monomolecular layers by means of the Langmuir balance makes it therefore possible to determine the molecular weight of the substance in the monolayer when the latter is in a highly diluted ("ideal" state), as well as the constants which characterize the intermolecular forces at higher concentrations ("real" state). Purely geometrical calculations permit valuable informa-

* Constants which characterise the so-called "volume" of molecules express in reality the effect of the short-distance intermolecular repulsion forces.

tion to be derived regarding the shape and size of molecules which form the monolayer, from the knowledge of the thickness of such a layer.

The simplest and most universal case occurs when the layer is highly diluted, so that the area $1/\Gamma$, assigned on the average to one molecule of the applied substance, is large compared with the area occupied by the molecule itself. In this case surface pressure is determined only by thermal motion, and there exists for it a valid ideal equation of state which, in this particular case, has the form

$$\pi = \Gamma kT ,$$

where Γ coincides with the previously given definition of this magnitude as the excess of the adsorbed component per unit surface. Therefore by measuring π one can determine the number of molecules in 1 cm^2, and, knowing the quantity of the applied substance, it is possible to calculate its molecular weight by analogy with the determination of molecular weight from osmotic pressure. The difficulties which the experimenter has to face consist mainly of the need for working with extremely diluted solutions (small values of Γ), where π is very low. This is particularly true in the most important cases (investigations of high-molecular compounds) for the same reasons which make the measurement of the osmotic pressures of the solutions of high polymers a difficult problem.

As surface concentration increases, the forces of molecular interaction begin to manifest themselves. Various cases are possible, some of them differing substantially from the interaction in the bulk (three-dimensional) phases. The simplest case occurs when, given a certain degree of packing in the monolayer, we obtain two two-dimensional phases: one with dense packing and the other as a two-dimensional solution. Any further increase of density will result in the increase of the area occupied by the condensed two-dimensional phase, whereas the area occupied by the two-dimensional solution will decrease, the surface tension remaining constant. This case resembles crystallisation from a solution at a constant concentration, or the condensation (or crystallisation) from the vapour phase at constant pressure. At high values of $1/\Gamma$ we obtain in the coordinates π, $1/\Gamma$, an isotherm having the form of a hyperbolic section which, at the abscissa corresponding to the beginning of "condensation", passes into a horizontal section corresponding to the heterogeneous monolayer.

On increasing the density further, when the two-dimensional gas has disappeared and the entire monolayer becomes again homogeneous, the

References p. 139–140

pressure will increase again, but much more steeply, owing mainly to intermolecular repulsion forces. This point is very distinct and allows a determination of Γ_∞ (or $1/\Gamma_\infty$) for the dense surface monolayer. This most pronounced "discontinuity" on the curve, which can be easily detected even by means of methods of poor sensitivity, was first noticed by Pockels and by Rayleigh. Fig. 32 shows several curves of this nature, obtained for penta-, tetra- and tridecanoic acids on water at 14.5° C.

The magnitude of the area $1/\Gamma_\infty$ occupied by a simple molecule in the dense monolayer provides valuable information on the position of this molecule in the dense monolayer and on its form. All monobasic higher fatty acids possess roughly the same values of $1/\Gamma_\infty$ ($\sim$ 20 Å^2) which is only possible if the molecules in question have an elongated shape with an identical cross section, and if they are arranged in a dense monolayer at right angles to the surface of water. The magnitude 20 Å^2 corresponds to the cross section of this molecule or, more exactly, to the section of the polar group COOH, which occurs in all these acids. The fact that these molecules "stand up" in a dense layer is readily explained by the great affinity of the COOH group towards water, whilst the paraffinic parts of the acid are only weakly linked. If the volume of the entire acid molecule is known (from its density), it is also easy to calculate its length, *i.e.* the

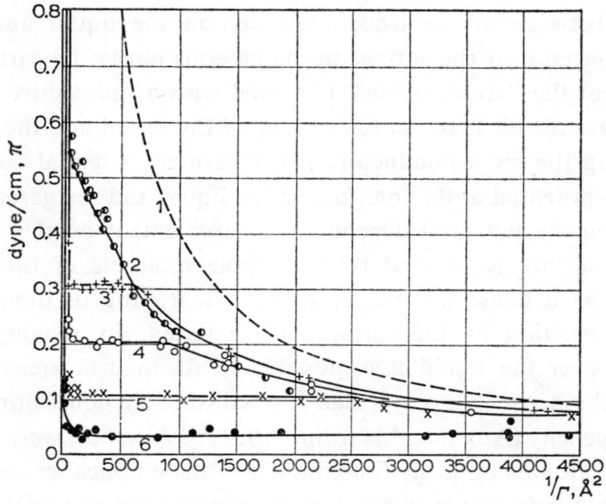

Fig. 32. Surface pressure/area-per-molecule isotherms for the lauric (2), tridecanoic (3), tetradecanoic (4), pentadecanoic (5) and palmitic (6) acid. For (3), (4) and (5) at 14.5°C; for (2) at 14–16°C; for (6) at 12°C.

distance between the carbon atoms in the hydrocarbon chain. The data thus obtained agree with those obtained by X-ray methods. The results obtained from direct measurements of surface pressure with the Langmuir balance agree perfectly with the concepts of the structure of the surface layer, indicated by surface tension measurements of the lower acids. When these results relate to comparatively simple substances whose structure has been well studied, they serve rather as an illustration and point to the possibilities offered by this method. The application of the method to more complex substances, for which other methods of structural research are rather ineffective, is of great interest.

The curves $\pi = \pi(1/\Gamma)$ shown in Fig. 32, resemble closely the $p = p(v)$ curves for the region which encompasses the gaseous, heterogeneous and condensed states. Many approximate formulae have been suggested for the $\pi = \pi(1/\Gamma)$ curves, which are of the van der Waals' equation type.

In a number of cases the isotherms $\pi = \pi(1/\Gamma)$ differ from the isotherms which had been analysed thus far, in that they have no horizontal part, but instead a gradual rise of π is observed up to the formation of the dense layer. This means that for such substances, *e.g.* lauric acid, no heterogeneous monolayer is formed and the packing proceeds gradually. The occurrence of the heterogeneous zone in the former case and its absence in the latter can be demonstrated directly, *e.g.* by measuring the potential difference between an electrode immersed in the liquid and another electrode placed near the surface in the gaseous phase. To carry out this measurement the "air electrode" is coated with a radioactive substance which ionizes the air between the surface of the liquid and the electrode, thus making the air a conductor. The measured potential comprises a drop in the potential at the interface of the liquid and the gaseous phase. This drop in the potential depends to a large extent on the manner in which the surface is covered by the dipole molecule of the insoluble substance: as a dense monolayer or as a scattering of dipoles in the horizontal position on the surface of the liquid. By moving the "air electrode" over the liquid it is possible to distinguish areas at which islands of the dense monolayer had formed, thus demonstrating directly its inhomogeneity (Adam and Harding, 1932)[71]. Attempts were also made to demonstrate this optically (making use of the elliptical polarization of light reflected from the surface), but the results obtained in this way do not appear to be very reliable.

The transition to the condensed state in monomolecular layers can

References p. 139–140

thus occasionally occur gradually. This gradual transition does not, so far, possess a three-dimensional analogue. The monomolecular films in this transition region are called liquid expanded films. The possibility of attaining this peculiar state is due to a supplementary orientating effect of the substrate on the adsorbed molecules. A sudden change of properties, *e.g.* between the gaseous and the liquid or crystalline state, is caused by the fact that the intermolecular forces, which vary very strongly with distance, cause the thermal motion of closely adjacent molecules in the condensed phase to acquire a character different from that of molecular motion in the interior of the gaseous phase. In the monomolecular layer on the liquid, as in adsorption layers in general, this difference of properties between the closely adjacent and the more widely spaced adsorbed molecules is considerably smaller, since both types interact with the base. When this interaction is sufficiently strong, the translational movement characteristic of non-condensed layers is rendered much more difficult and the changes during condensation are therefore reduced or, more exactly, the formation of condensed islands in the monolayer becomes improbable, so that the transition occurs gradually.

During the investigation of the relation $\pi = \pi(1/\Gamma)$, cases intermediate between the above two extremes analysed so far were also observed, in which the discontinuity corresponding to the beginning of two-dimensional condensation does occur, but the curve is not parallel to the abscissa up to the second point of inflexion (dense layer) but rises instead, and the transition to the dense layer takes place very smoothly (*e.g.* for ethyl palmitates).

Hundreds of substances were studied by the above methods, and many interesting results were obtained. As was already mentioned, investigations of the more complex high-molecular substances are of particular importance. It should however be kept in mind that the observed properties of surface films are not identical with the properties of the same substances in the three-dimensional state. Moreover, the role played by the base is very great, and its importance, and consequently the characteristics of the "monomolecular" state, is manifested to a greater extent in the case of high-molecular substances with a more complicated structure.

The work on egg albumin can be quoted as an example of such investigations. This is a relatively simple protein, capable of being thoroughly purified. The first measurements on monolayers of egg albumin and other proteins, including the determination of the molecular weight, were carried out by Guastalla in 1945[72]. Using a surface balance of special construction, with an accuracy of 0.001 dynes (!), this author demonstrated that it is only at very high dilutions of the surface layer (100 m^2/mg) that the dependence of the two-dimensional osmotic pressure on area is of the same nature as in ideal solutions. Under these conditions the molecular weight of egg albumin was found to be 40,000. Using a different solution as a base (a concentrated aqueous solution of ammonium sulphate instead of Guastalla's acidified aqueous solution), Bull (1947)[73] established the existence of a wider region of the ideal two-dimensional state (up to 1.5 m^2/mg). Working with greater accuracy than Guastalla, Bull checked Guastalla's results and found that $M = 44{,}000$. In 1951, Mishuck[74], using a more convenient method of measurement, found $M = 44{,}900$ for very carefully purified egg albumin, using a concentrated solution of ammonium carbonate as the base. The area-dependence of π (in m^2 per 1 mg of applied substance) obtained by Mishuck is shown in Fig. 33.

The results of these investigations included also the minimum area

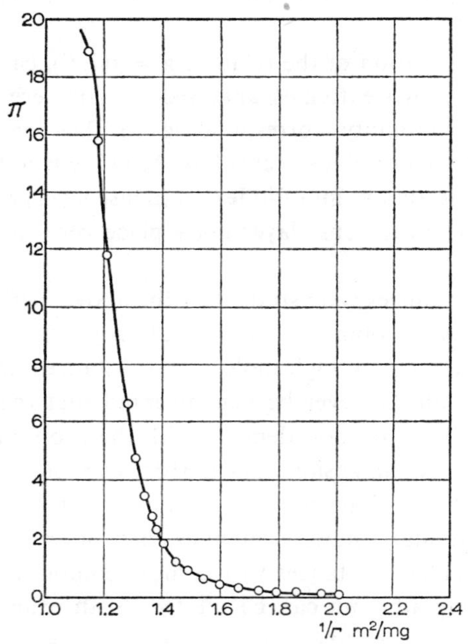

Fig. 33. Surface-pressure isotherm for albumin.

References p. 139–140

which still corresponds to the "gaseous" state, the area/molecule at which formation of condensed monolayers begins, and other constants which can help us in drawing important conclusions with respect to the structure of the albumin molecule and to its position in the monolayer.

REFERENCES

1 FREUNDLICH, *Kapillarchemie*, 4. Aufl., Hirschwald, Berlin, 1930.
2 ADAM, *The physics and chemistry of surfaces*, 3rd ed., Oxford University Press, London, 1941.
3 ADAMSON, *Physical chemistry of surfaces*, Interscience, New York, 1960.
4 GIBBS, *The collected works*, Longmans, Green and Co., New York, 1928, Vol. 1.
5 LAPLACE, *Oeuvres complètes*, Paris, 1806, Tome 4.
6 SHCHERBAKOV, *Kolloidn. Zh.*, 20 (1958) 502.
7 MANDELSTAMM, *Ann. Phys. (Paris)*, 41 (1913) 609.
8 VOLMER, *Z. Physik. Chem. (Leipzig)*, 125 (1926) 151; 206 (1957) 181.
9 REHBINDER, *Z. Physik*, 72 (1931) 191.
10 SHCHUKIN AND REBINDER, *Kolloidn. Zh.*, 20 (1958) 645.
11 POLANYI, *Ber. Deut. Physik. Ges.*, 16 (1914) 1012.
12 HÜCKEL, *Adsorption und Kapillarkondensation*, Akademischer Verlagsgesellschaft, Leipzig, 1928.
13 HAMAKER, *Physica*, 4 (1937) 1058.
14 DERYAGIN, *Zhur. Fiz. Khim.*, 5 (1934) 379.
15 MARTYNOV AND DERYAGIN, *V konferencija po kolloidnoj khimii, tezisy (5th conference on colloid chemistry, theorems)* Moskva, 1962.
16 SHCHERBAKOV, *Kolloidn. Zh.*, 20 (1958) 759.
17 SHCHERBAKOV, *Izvest. VUZ-fizika*, 4 (1959) 77.
18 SHCHERBAKOV AND RYANTSEV, *Zhur. Fiz. Khim.*, 34 (1960) 2120.
19 FRUMKIN, *Zhur. Fiz. Khim.*, 12 (1938) 337.
20 TOLMAN, *J. Chem. Phys.*, 17 (1949) 333.
21 BUFF, *J. Chem. Phys.*, 19 (1951) 1591.
22 HILL, *J. Phys. & Colloid. Chem.*, 54 (1950).
23 VOLMER, *Kinetik der Phasenbildung*, Dresden, 1939.
24 FARKAS, *Z. Physik. Chem.*, 125 (1927) 236.
25 KAISHEV AND STRANSKI, *Z. Physik. Chem. (Leipzig)*, B 26 (1934) 317.
26 BECKER AND DÖRING, *Ann. Physik*, 24 (1935) 719.
27 FRENKEL, *Fiziol. Zhur. SSSR*, 1 (1939) 243.
28 ZEL'DOVICH, *Acta Physicochim., URSS*, 18 (1943) 1.
29 VOLMER AND WEBER, *Z. Physik. Chem.*, 119 (1925) 227.
30 VOLMER AND FLOOD, *Z. Physik. Chem. (Leipzig)*, A 170 (1934) 273.
31 SANDER AND DAMKÖHLER, *Naturwissenschaften*, 31 (1943) 460.
32 CLARKE, Thesis, Urbana, 1953.
33 THOMSON, *Elektrizitätsdurchgang in Gasen*, Leipzig, 1906.
34 TOHMFOR AND VOLMER, *Ann. Physik*, 33 (1938) 109.
35 LIPPMANN, *Pogg. ann.*, 149 (1873) 547.
36 KRĂSTANOV, *Trudove na CMI (Bulgaria)*, 1 (1941) 15.
37 KRĂSTANOV, *Időjárás*, 61 (1957) 53.
38 KAISHEV AND MUTAFCHIEV, *Izv. Bulgar. Akad. Nauk*, 7 (1960) 177.
39 MUTAFCHIEV AND TOSHEV, *Izv. Inst. Fiziko-khim. Bulgar. Akad. Nauk*, 1 (1960) 43.

40 Matijevič and Ottewill, *J. Colloid Sci.*, 13 (1958) 342; *Kolloid-Z.*, 161 (1958) 50.
41 Shcherbakov, *Kolloidn. Zh.*, 20 (1958) 759; Shcherbakov, Ryantsev and Filippova, *Kolloidn. Zh.*, 23 (1961) 338; Shcherbakov, *Kolloidn. Zh.*, 24 (1962) 502.
42 Nikolov, *Annuaire Univ. Sofia*, 54 (1959/60) 203.
43 Knudsen, *Ann. Physik (Paris)*, 47 (1915) 697.
44 Langmuir, *J. Am. Chem. Soc.*, 38 (1916) 2221.
45 Szyszkowski, *Z. Physik. Chem.*, 64 (1908) 385.
46 Taubman, *Doklady Akad. Nauk SSSR*, 71 (1950) 343.
47 Duclos, *Ann. Phys. Chim. (Paris)*, 13 (1878) 76; Traube, *Ann.*, 265 (1891) 27.
48 Frumkin, *Z. Physik. Chem.*, 116 (1925) 466.
49 Volmer and Mahnert, *Z. Physik. Chem.*, 115 (1925) 239.
50 Cassel and Formstecher, *Kolloid-Z.*, 61 (1932) 18.
51 Ekserova and Sheludko, *Izv. Inst. Fiziko-khim., Bulgar. Akad. Nauk*, 3 (1963) 79; Sheludko, Ekserova and Platikanov, *Kolloidn. Zh.*, 25 (1963) 606.
52 Rayleigh, *Proc. Roy. Soc. (London)*, A92 (1915) 184.
53 Bashforth and Adam, *An attempt to test the theories of capillary action*, Cambridge, 1883.
54 Sugden, *J. Chem. Soc.*, 119 (1921) 1483.
55 Harkins and Brown, *J. Am. Chem. Soc.*, 41 (1919) 499.
56 Sugden, *J. Chem. Soc.*, 121 (1922) 858; 125 (1924) 27.
57 Warren, *Phil. Mag.*, 4 (1927) 358.
58 Rebinder, *V cĕezd russkih fizikov (5th congress of Russian physicists)* Moskva, 1926, p. 72.
59 Tate, *Phil. mag.*, 27 (1864) 176.
60 Lottermoser, *Kolloid-Z.*, 73 (1935) 155.
61 Du Nouÿ, *Surface equilibria of colloids*, Chem. Catalog Co., 1926.
62 Harkins and Jordan, *J. Am. Chem. Soc.*, 52 (1930) 1751.
63 Nutting and Long, *J. Am. Chem. Soc.*, 63 (1941) 84.
64 Smolders, Thesis, Utrecht, 1961.
65 Levich, *Physicochemical Hydrodynamics*, Prentice-Hall, Englewood Cliffs, 1962.
66 Pockels, *Nature*, 43 (1891) 437.
67 Rayleigh, *Phil. Mag.*, 48 (1899) 337.
68 Langmuir, *J. Am. Chem. Soc.*, 39 (1917) 1848.
69 Adam and Jessop, *Proc. Roy. Soc. (London)*, 110 A (1926) 423.
70 *Monomolekulyarnye sloi, Sbornik statei*, Inostrannaya literatura, Moskva, 1956.
71 Adam and Harding, *Proc. Roy. Soc. (London)*, 138 A (1932) 411.
72 Guastalla, *Research (London), Suppl. Surface Chem.*, 145 (1945).
73 Bull, *Advances in Protein Chem.*, 3 (1947) 95.
74 Mishuck and Eirich, *J. Polymer Sci.*, 7 (1951) 341.

CHAPTER V

Electrokinetic Phenomena

Phenomena in which an electric field produces a transport of the disperse phase or the dispersion medium, or else in which an electric field arises as a result of the transport of one phase within another, are called electrokinetic phenomena. Four types of such phenomena are known.

Electro-osmosis consists in the movement of a liquid through a capillary or a porous diaphragm under the influence of an electric field. This phenomenon was discovered by Reuss (1809)[1] and investigated quantitatively by Wiedemann (1852)[2] and Quincke (1861)[3]. An apparatus for its measurement is shown diagrammatically in Fig. 34. When a potential difference is applied to electrodes E_1 and E_2, arranged on both sides of the porous diaphragm D (or of a capillary), the aqueous solution begins to flow through D, displacing the meniscus M along the capillary C. Wiedemann established that the quantity of the liquid transferred is proportional to the current between E_1 and E_2, and that for a constant current intensity it is independent of the area and thickness of the porous diaphragm.

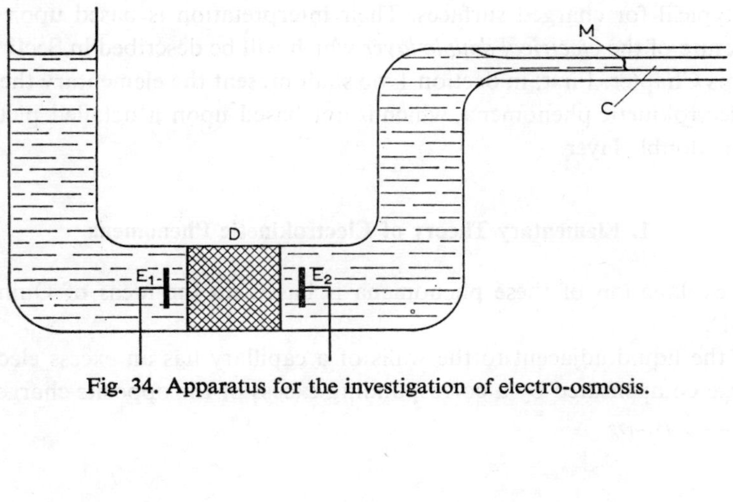

Fig. 34. Apparatus for the investigation of electro-osmosis.

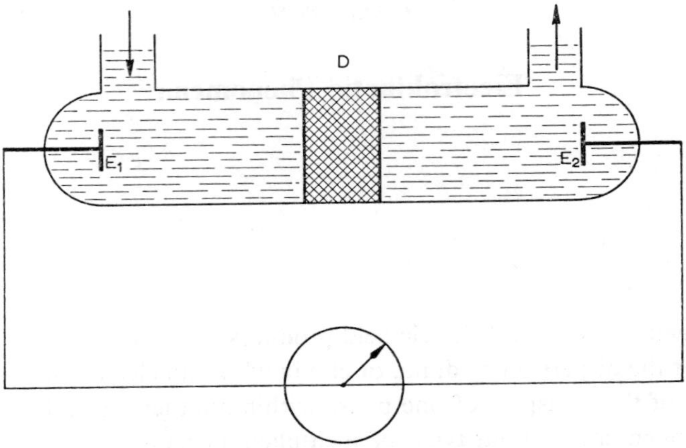

Fig. 35. Apparatus for the investigation of the streaming potential.

The *streaming potential*, which can arise when a liquid flows through a capillary or a porous diaphragm was discovered by Quincke in 1859[4] (Fig. 35). This author also found that the generated potential difference is directly proportional to the pressure difference which causes the flow, and is independent of the area and thickness of the porous diaphragm.

Electrophoresis, *i.e.* the transfer of colloidal particles in an electric field, was discovered by Reuss simultaneously with electro-osmosis.

The *sedimentation potential*, which arises along the direction of sedimentation during the sedimentation of particles, was described by Dorn in 1880[5].

As was demonstrated in later investigations, electrokinetic phenomena are typical for charged surfaces. Their interpretation is based upon the structure of the *electrical double layer* which will be described in Section 2 of this Chapter. First, in Section 1 we shall present the elementary theory of electrokinetic phenomena, which is not based upon a detailed picture of the double layer.

1. Elementary Theory of Electrokinetic Phenomena

The explanation of these phenomena is based on the ideas of Quincke (1861).

If the liquid adjacent to the walls of a capillary has an excess electric charge compensated by a corresponding excess of the opposite charge on

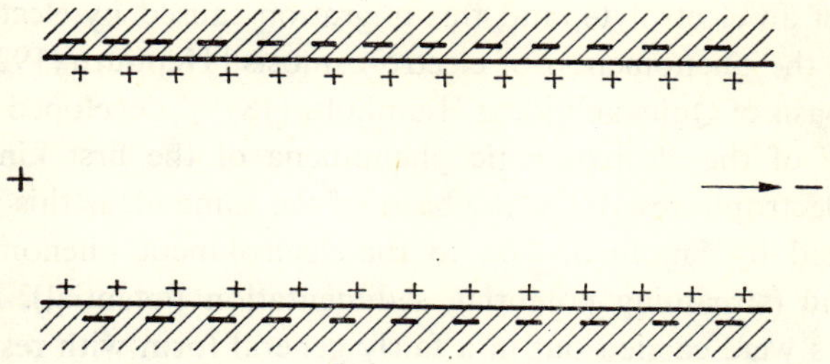

Fig. 36. Electrical double layer on the walls of a capillary.

the wall (Fig. 36), then, if an electric field is applied along the capillary, a force will arise which will tend to displace the charges of the liquid (and consequently also the liquid itself) within the capillary with respect to its walls, thus bringing about the electro-osmosis. If on the other hand we cause the liquid flow through the capillary by generating a pressure difference between the two ends, the charges of the liquid will be displaced along the capillary and give rise to the streaming potential. The existence of a charge on the surface of colloidal particles with respect to the liquid phase will cause these particles, in the same way as ions, to migrate in the electric field (electrophoresis). Finally, if charged particles settle, this will be associated with a transfer of their charge along the direction of sedimentation, and with a convective sedimentation current and the corresponding electric field—the sedimentation potential.

In addition to this explanation of electrokinetic phenomena, from the point of view of the physical chemistry of surfaces, Wiedemann (1856)[6] also suggested another, "bulk" explanation of electro-osmosis, according to which the liquid is carried away by ions which move in its bulk due to an electric field. Since the mobilities (friction of ions against the liquid) of oppositely charged ions are generally unequal, it is possible for the transport of the liquid by cations in one direction to differ from the transport of the liquid by anions in the opposite direction; this may give rise to a flow or migration in the solution. The velocity of this migration should however be independent of the nature of the wall and correspondingly of the structure of the wall/liquid interface and of the section of the capillary. It will be demonstrated below that electrokinetic phenomena are greatly influenced by the charge of the phase surface, and are only observed (in the case of electro-osmosis and streaming potential) in narrow capillaries. The "bulk" transfer of the liquid takes place in concentrated solutions of an electrolyte, and this migration caused by electrolysis can complicate the phenomenon of electro-osmosis [Hepburn (1927–1933)[7]].

On the basis of Quincke's ideas Helmholtz (1879)[8] developed a quantitative theory of the electrokinetic phenomena of the first kind (electro-osmosis, electrophoresis). On the basis of the same ideas this theory was also adapted by Smoluchowski to the electrokinetic phenomena of the second kind (streaming potential, sedimentation potential). Helmholtz' calculations were carried out in a fairly general form with respect to the structure of the double layer, but his final results are simplified by the assumption that the double layer is very thin. Introducing such a simplification right from the outset, Perrin (1904)[9] put forward a very graphic method of calculation which yields identical results; this method will be used in the present book. The review by Smoluchowski[10] provides more detailed information regarding these problems.

Let us first assume that the surface charge ($-e$ per cm^2) is compensated by a liquid charge ($+e$* per cm^2) at a very short distance δ. This is electrically equivalent to a condenser with charge e, plate distance δ and potential

$$\zeta = \frac{4\pi\delta e}{\varepsilon}, \tag{5.1}$$

where ε is the dielectric constant of the material in the condenser.

In the electric field of strength E directed parallel to the interface, each cm^2 of the surface of the liquid will be acted upon by the force

$$f_1 = Ee, \tag{5.2}$$

which tends to move the liquid with respect to the wall. When the force f_1 which causes this migration becomes equal to the frictional force f_2 per cm^2 between the liquid and the wall, which according to Newton's law, is

$$f_2 = \eta \frac{u}{\delta} \tag{5.3}$$

(where η is the viscosity of the liquid in the double layer, and u/δ is the

* This "e" must not be confused with the proton charge, which is denoted by the same letter.

References p. 171–172

velocity gradient in the direction at right angles to the migration), a constant velocity of migration, *i.e.* a constant velocity of the liquid in the capillary will be established. By equating f_1 with f_2 we obtain for the steady-state flow

$$Ee = \eta \frac{u}{\delta} \tag{5.4}$$

or, according to (5.1), for the rate of flow

$$u = \frac{\varepsilon\zeta}{4\pi\eta} E. \tag{5.5}$$

The amount v of liquid transferred through a capillary of radius r during 1 sec is therefore*

$$v = u\pi r^2 \tag{5.6}$$

so that according to (5.4)

$$v = \frac{\pi r^2 \delta e}{\eta} E \tag{5.7}$$

or, making use of (5.1),

$$v = \frac{\varepsilon\zeta r^2}{4\eta} E. \tag{5.7'}$$

It is usually more convenient to investigate electro-osmosis not in a single capillary with a cross section πr^2, but in a porous diaphragm or plug (of thickness l) with a large number of capillaries. In this case the effect (v) is considerably greater, since electro-osmosis occurs simultaneously through a large number of capillaries whose total cross section is $\sum \pi r^2$. For a porous diaphragm made of non-conducting material the total cross section can easily be determined from the electric resistance. If the capillaries are filled with a solution of specific conductivity κ:

$$R = \frac{1}{\kappa} \frac{l}{\sum \pi r^2}. \tag{5.8}$$

On the other hand, according to Ohm's law, $R = El/i$, where El is the

* Provided $r \gg \delta$.

potential difference between both sides of the diaphragm, and i is the current intensity, Therefore

$$\sum \pi r^2 = \frac{i}{\kappa E}, \tag{5.9}$$

and after substituting $\sum \pi r^2$ for πr^2 in (5.7) and (5.7′), we obtain the convenient form

$$v = \frac{\delta e}{\eta \kappa} i = \frac{\varepsilon \zeta}{4\pi \eta \kappa} i. \tag{5.10}$$

When using formula (5.10) one should take into consideration the fact that ε, η and κ are not necessarily the same in this case as in a larger mass of the liquid. For instance the conductance near the walls of the capillary is, even if only because of the excess of e-determining ions*, not the same as κ_∞ inside the solution. Having introduced the concept of surface conductance κ_s, Smoluchowski[10] put forward the following formula for capillaries which are not extremely narrow**:

$$\kappa = \kappa_\infty + \frac{S}{O} \kappa_s, \tag{5.11}$$

where S is the perimeter of the cross-section O of the capillary. For a cylindrical capillary $S/O = 2\pi r/\pi r^2 = 2/r$, so that the correction for the surface conductance $(\kappa - \kappa_\infty)$ decreases with increasing r***

* Charge-determining ions, which are responsible for the charge on the capillary wall.

** Theoretical explanations of the surface conductance [Bikerman (1935)[11], Urban, White and Strassner (1935)[12]] take into account not only the concentration of charges at the interface, but also their electro-osmotic displacement.

*** Smoluchowski's equation (5.11) is derived on the basis of the conventional concept of the resistance of parallel conductors a surface layer with a specific surface conductance κ_s and an inner cylinder of the liquid with a specific conductance κ_∞. The conductance of the whole system is then:

$$1/R = 1/R_\infty + 1/R_s,$$

where $1/R = \kappa O/l$, $1/R_\infty = \kappa_\infty O/l$ and $1/R_s = \kappa_s S/l$, which is permissible if the thickness of surface layer is small compared with the capillary radius, and yields (5.11). The quantity κ_s is determined experimentally on the basis of (5.11). It should be noted that surface conductance is not necessarily the only cause of the dependence of, say, v/i (5.10) on the capillary radius.

Note should be made of the so-called capillary superconductivity, discovered by Friedrichsberg in 1947, for which the surface conductance is not the correction but the basic, determining factor. This effect consists in that a porous body made of a non-conducting material, which is filled with a solution of an electrolyte, sometimes exhibits a lower resistance than a solution of the same cross-section. In this case surface conductance obviously overcompensates the reduction of conductance due to the non-conducting skeleton of the porous body.

References p. 171–172

Eq. (5.10) may be applied to any arrangement of capillaries in a porous diaphragm, since both the electric current and the electro-osmotic stream flow parallel to the capillary walls[13].

A special case of electro-osmosis is the so-called electro-osmotic pressure. When the capillary C (Fig. 34) is not horizontal but either inclined or vertical, then the liquid level in it varies during the electro-osmotic process, and a hydrostatic pressure is formed, which counteracts the electro-osmotic flow. The meniscus level finally settles in a certain constant position which corresponds to a constant drop of hydrostatic pressure, the electro-osmotic pressure ΔP. The state thus obtained is outwardly similar to the equilibrium in an osmotic cell. In fact it has nothing in common with an equilibrium, but is a steady state; this fact is not taken into account by many investigators and gives rise sometimes to gross errors. When the steady state meniscus level has been established in the capillaries of a porous diaphragm, electro-osmotic transfer continues near the walls, owing to the application of the electric field. A reverse flow due to hydrostatic pressure occurs simultaneously within the capillaries. This steady state corresponds therefore not to the equality of electro-osmotic and hydrostatic pressures, but to the equality of the amounts of liquid, transferred during 1 sec by electro-osmosis (4.7′) and due to the hydrostatic pressure according to Poiseuille's law $\Delta P \pi r^4/8\eta l$. The electro-osmotic pressure obtained (at $\delta \ll r$) is therefore

$$\Delta P = \frac{2\varepsilon\zeta}{\pi r^2} El. \tag{5.12}$$

Eqn. (5.10) agrees with the independence of v of the thickness and cross-section of the porous diaphragm established by Wiedemann and with its proportionality to i, whilst (5.12) is borne out by the proportionality between ΔP and E/r^2, established by Wiedemann and Quincke.

Both formulae demonstrate a close relation between electro-osmosis and the structure of the phase surface, characterised by the electric parameters $e\delta$, or $\varepsilon\zeta/4\pi$.

The velocity of electrophoretic migration can be readily calculated on the strength of the considerations referred to above. The mechanism of electrophoresis is analogous to that of electro-osmosis, the difference being that, in the former case, the liquid phase remains stationary and the particles are displaced, owing to their small size and consequent easy mobility. The velocity of their migration can be found by means of (5.4).

If this velocity u is divided by the strength of the applied electric field E, one obtains the magnitude $u_e = u/E$, which is known as electrophoretic mobility. This quantity characterises the mobility of a colloidal particle during electrophoresis. The magnitude u_e is completely analogous to the ionic mobility in an electric field. According to (5.4) and (5.1)

$$u_e = \frac{e\delta}{\eta} = \frac{\varepsilon\zeta}{4\pi\eta}. \tag{5.13}$$

The above equation is not fully exact. In 1931 Henry carried out a more detailed analysis of this phenomenon by making use of the ideas of the Debye–Hückel theory relating to ionic atmospheres and their effect on ionic mobilities, finding that

$$u_e = f\frac{\varepsilon\zeta}{\pi\eta}, \tag{5.14}$$

where $f = f(\varkappa, r)$ depends on the thickness of the ionic atmosphere $1/\varkappa$ (see next paragraph) around the particle and on its radius r. For cylindrical particles whose axis coincides with the direction of the field $f = \frac{1}{4}$, just as in formula (5.13); for cylindrical particles whose axis is at right angles to the field, $f = \frac{1}{8}$ at small $\varkappa r$ and approaches $\frac{1}{4}$ at larger values of $\varkappa r$; for spherical particles $f = \frac{1}{6}$ at small $\varkappa r$, and $\frac{1}{4}$ at large $\varkappa r$. This is valid for non-conducting particles. If the particles consist of a conductor, *e.g.* a metal, the above considerations are only valid for small values of $\varkappa r$. For small ionic atmospheres and large particles, *i.e.* for large $\varkappa r$, the mobility of conducting particles tends to zero. In other words, according to this theory the electrophoretic mobility of colloidal particles depends, through f, on their dimensions.

This conclusion and other considerations which also predict that the radius should have an effect on mobility (effect of the relaxation of ionic atmopheres) cannot be accepted as completely confirmed experimentally. From the example of suspensions of quartz particles and particles of other substances, coated by adsorption with egg albumin, Freundlich and Abramson (1927–1928)[14] demonstrated that the electrophoretic mobility of these particles does not depend on their dimensions. Since the particles in question were large (over 1 μ), whereas the thickness of ionic atmosphere $1/\varkappa$ is small (less than 10^{-6} cm), the condition of $\varkappa r \gg 1$ had been complied with and the lack of dependence between u_e and r can be explained. However, Overbeek (1950)[15] found that the mobility of egg albumin molecules ($r = 2 \cdot 10^{-7}$ cm) is the same as that of large particles coated with egg

References p. 171–174

albumin, and this is at variance with the requirements of the theory. On the other hand Mooney (1924)[16] established that the electrophoretic mobility of oil droplets depends on their size.

The fact that the theory has not as yet been checked accurately by experiment prevents definite conclusions, relating to the ζ-potential (or charge) of colloidal particles, from being drawn on the basis of their electrophoretic mobility. Nevertheless, in view of its simplicity as a method of identification, electrophoresis has found wide application particularly for biological systems. It is therefore proposed to analyse below in more detail the methods by which the electrophoretic mobility may be determined.

Electrokinetic phenomena of the second kind—streaming and sedimentation potentials—can also (as mentioned already) be analysed on the basis of the concept of the electrical double layer. Let us first analyse the former case.

Let us assume that the pressure gradient along a capillary filled with a liquid is dP/dl, where l is the distance from the end of the capillary. According to Poiseuille's law, this will result in a flow having maximum velocity along the capillary axis, the velocity decreasing with a parabolic profile, to the value of zero at the walls. At a distance $x < r$ from the capillary axis

$$u_x = \frac{1}{4\eta}\frac{dP}{dl}(r^2 - x^2),$$

or, for a cylindrical capillary of constant section ($dP/dl = P/l$, where l is the length of the capillary and, P is the pressure drop between its ends), we have at a small distance δ from the wall

$$u_\delta = \frac{1}{4\eta}\frac{P}{l}[r^2 - (r-\delta)^2] \approx \frac{1}{2\eta}\frac{P}{l}r\delta \tag{5.15}$$

for $\delta \ll r$. If at a distance δ there is an excess charge e per 1 cm^2, this charge will be transported along the wall at the same speed, resulting in the convection current

$$i_c = 2\pi r u_\delta e\,. \tag{5.16}$$

As these charges are transferred from one end of the capillary to the other, they generate a potential difference El which increases with increasing accumulation of charges and causes an increasing conduction current i

through the solution in the capillary of resistance R. When this conduction current becomes equal to the convection current, a constant, steady-state value of $El = \Delta V_{st}$ will be established, where ΔV_{st} is the streaming potential. According to Ohm's law

$$i = \frac{\Delta V_{st}}{R} \quad \text{at} \quad R = \frac{1}{\kappa}\frac{l}{\pi r^2},$$

which, after equating i_c (5.16) with i, and taking into account (5.15), yields

$$\Delta V_{st} = \frac{\delta e}{\eta\kappa}P = \frac{\varepsilon\zeta}{4\pi\eta\kappa}P, \tag{5.17}$$

the latter in accordance with (5.1).

Eq. (5.17) remains valid in the case of a porous diaphragm independently of the form of its capillaries, for reasons given in connection with (5.12); the notes relating to the dependence of κ on pore or capillary radius remain in force.

The sedimentation potential E_{sed}, *i.e.* the potential difference arising during sedimentation between electrodes separated vertically by a distance of 1 cm, is calculated in a similar way.

The sedimentation rate of spherical particles is given by (3.22) as $u = (2r^2/9\eta)(\rho - \rho_0)g$.

If the charge of a colloidal particle is q and the number of such particles in 1 cm^3 is c, then the convection current density (current per 1 cm^2 of cross-section normal to the direction of sedimentation) will be

$$I_c = quc. \tag{5.18}$$

In the steady state this density is equal to the density of the conduction current $I = E_{sed}l/R$, where $E_{sed}l$ is the steady-state potential difference between electrodes separated vertically by a distance l.

The relation between the charge of a spherical particle q and the zeta potential can be found from the condition of equality between the electric force qE acting on the particle during electrophoresis and the frictional force $u_eE6\pi\eta r$ which arises when the particle moves at a speed u_eE, where $u_e = \varepsilon\zeta/4\pi\eta$ in accordance with (5.13):

$$q = \tfrac{3}{2}\varepsilon\zeta r. \tag{5.19}$$

By equating I with I_c we obtain

$$E_{sed}\, l/R = quc\,,$$

and after substituting the found values of u, q and $R = l/\kappa$ (unit cross section), we obtain

$$E_{sed} = \frac{\varepsilon\zeta}{3\eta\kappa} r^3(\rho - \rho_0)cg\,. \tag{5.20}$$

Eq. (5.20) was checked experimentally by Stock (1913)[17].

By comparing the results for the streaming potential (5.17) with those for the electro-osmotic velocity (5.10) for the same system, we obtain

$$\frac{\Delta V_{st}}{P} = \frac{v}{i}\,. \tag{5.21}$$

This relation is sometimes called Saxen's equation, after the latter author who found it experimentally in 1892[18]. Mazur and Overbeek (1951)[19] substantiated this relation on the basis of Onsager's general theory of irreversible processes[20].

The theory given above is only approximate for many reasons. It does not allow for the fact that the so-called electro-viscous drag is produced as a result of electrokinetic movement*. In 1955 Sheludko[21] showed that if this factor is allowed for, corrections should be introduced into all formulae of the elementary theory when the thickness of the electrical double layer is fairly large, which, as will be demonstrated, is frequently the case in practice.

An important supplement of the theory is found in papers by Deryagin and Dukhin (1959)[23]. These authors take into account the effect of ionic diffusion which is associated with the electrokinetic phenomena. This was found to be particularly essential for liquid interfaces, *e.g.* for the Dorn effect during the rise of gaseous bubbles. When a solid spherical particle

* The field generated during any migration of charges along the interface tends to counteract the migration. Electro-viscosity is usually strictly understood to be the increase in the viscosity of a suspension of charged particles in comparison with the viscosity of the same suspension, when the particle charge is equal to zero. For this case Smoluchowski suggested a formula, which was later substantiated theoretically by Krasny-Ergen (1936)[22]. The author demonstrated in 1955 that this formula is incorrect[21]. In view of the unclear nature of this phenomenon, no further discussion will be devoted to it.

moves in a solution of an electrolyte, a concentration difference between its poles along the direction of movement, and a corresponding diffusion potential, may also be observed. Allowance for this potential leads to a correction which can be of the same order as the displacement potential of the particle. When the theory is made more accurate by taking into account the diffusion and the preservation of anions and cations, equations are obtained, which assume the classical form only if the diffusion coefficients are identical for anions and cations. By allowing for the diffusion, it became possible to arrive at the conclusion (based on Onsager's ideas of the symmetry of kinetic coefficients) that a flow of the solution (capillary osmosis) should arise if a concentration difference exists between both sides of the capillary or porous diaphragm, and also that the suspended particles should move in the solution down the concentration gradient ("diffusion-phoresis"). The relevant calculations and equations will not be quoted here owing to the lack of space.

2. Structure of the Diffuse Double Layer

The elementary theory of electrokinetic phenomena characterises the electrical double layer by e and δ, or ζ and ε, these magnitudes being used in pairs in expressions for the velocity of electrokinetic phenomena or for the displacement potentials. For this reason, Guggenheim (1940)[24] suggested the use of these magnitudes as products $e\delta$ or $\varepsilon\zeta/4\pi$ to characterise the moment of the double layer, the unit of measurement being called a "helmholtz", after the author of the theory of electrokinetic phenomena.

Following more detailed study, the physical nature of these four magnitudes was found to be less simple than it seemed from Quincke's ideas. The electric field in any material medium, which ultimately consists of positively charged nuclei and negatively charged electrons, has an extremely complex structure determined by the spatial distribution and the magnitude of these charges. The mean potential φ of this field is therefore a function of the form and state of the phase, and is as a rule different for various phases. Hence a potential jump φ_0 occurs between two phases, equal to the difference between the potentials within the two adjacent phases. However, according to Quincke's ideas, this potential jump is localised at the interface where the accumulation of excess charges takes place. This form of the relationship $\varphi(x)$ (x being the direction normal to the plane phase surface) is not compulsory and does not necessarily follow

References p. 171–172

from the existence of the difference φ_0. On the contrary, one would be entitled to assume that the change of potential on passing from one phase to another would take place gradually, in conformity with the specific structure of the substance in the interface zone, and that it would not therefore be determined merely by the state in the bulk of the phases.

If the double layer was in the form of a molecular condenser, as in Quincke's scheme, then the potential φ_0 and the zeta potential should be equal. This however is not the case. It is very easy to establish experimentally the difference between φ_0 and the zeta potential for interfaces with which one can both carry out electrokinetic measurements and measure the potential φ_0 except for an additive constant. Such investigations were carried out by Freundlich and Ettisch (1925)[25] with glass. The zeta-potential calculated from the streaming potential for the flow through a glass capillary was found to be a completely different funcion of the concentration of various electrolytes than the electrochemical potential of a glass electrode of the same glass (Fig. 37). These data were later confirmed by Kruyt and van der Willigen (1928)[26], who determined the streaming potentials for quartz and glass. The same authors also observed a considerable lowering of the zeta-potential with increasing concentration of the electrolyte. Ions of higher valency exerted a stronger effect, while tri- and tetravalent ions may even reverse the sign of the zeta potential.

Analogous results for the zeta-potential at the glass/aqueous solution interface, which demonstrate the ζ-lowering effect of electrolytes and the greater effectiveness of the ions of higher valency, were also obtained

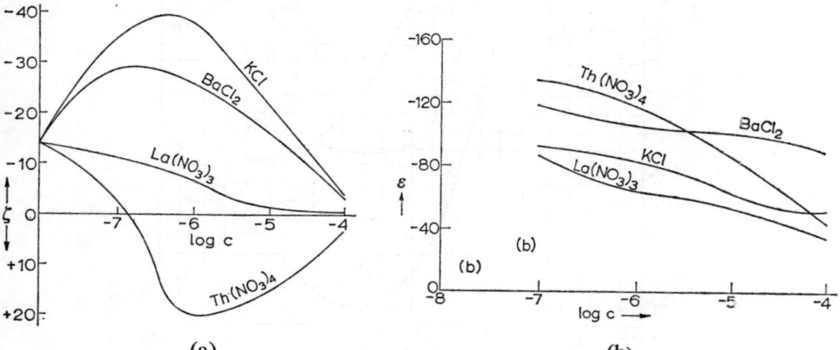

(a) (b)

Fig. 37 (a and b). Dependence of the ζ and ε potentials on log c (c mole/l; glass/aqueous solution of electrolyte).

more recently, *e.g.* by Wijga in 1946[28] (electro-osmosis and streaming potential). A very strong effect, with reversal of the sign of the zeta-potential by multivalent ions was described by Rutgers and De Smet in 1945[28] (Fig. 38).

Similar results were obtained for other interfaces. Thus from analytical determinations of the point of zero charge for silver halides, and calculations of the zeta-potential from electrophoresis, it was established that the zero charge points obtained by both methods coincide. The dependence of ζ on the potential-determining electrolyte concentration is, however, complex and not of the Nernst form [Troelstra and Kruyt (1942)[29]]. This is illustrated in Fig. 39.

These and many other experimental works show unequivocally that the φ_0 and ζ potentials are different magnitudes, which is incompatible with the ideas of the molecular structure of the electrical double layer. This discrepancy is partly explained by the theory of Gouy and Chapman, which is summarised below.

Gouy (1910)[30] and Chapman (1913)[31] pointed out, almost simultaneously and independently of one another, the important fact that an

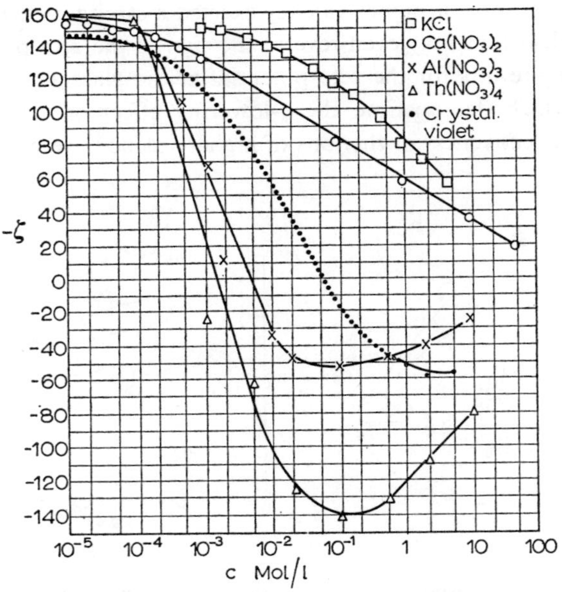

Fig. 38. Dependence of the ζ-potential on log c (glass/aqueous solution of electrolyte).

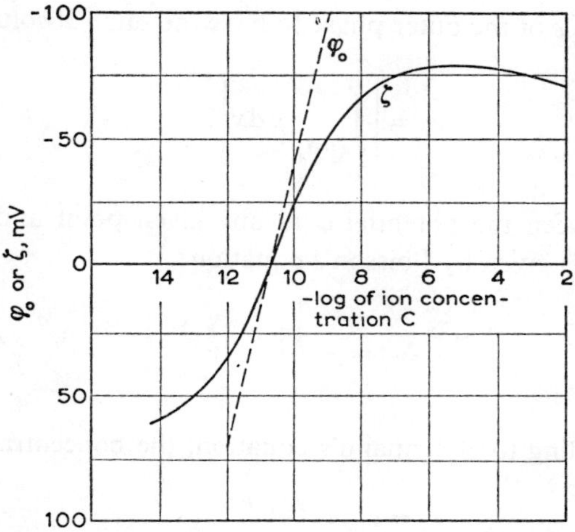

Fig. 39. Dependence of φ_0 (calculated) and of ζ-potential on log c (AgI/queous) solution of electrolyte).

excess of similarly charged ions cannot be wholly concentrated at the interface, because they are also under the influence of thermal motion which tends to disperse the ions evenly throughout the whole system. Consequently the concentrations of ions near the interface are determined by competition between the effects of the electric field and of the molecular kinetic motion. This diffuse distribution of ions is associated with a smooth variation of potential from its value φ_0 at the phase surface to zero in the bulk of the solution*.

The idea of the bulk structure of the electric layer in the solution presupposes utilisation of the concept of the space charge density. This density ρ can be expressed as

$$\rho = \sum F z_i c_i, \tag{5.22}$$

where z_i is the valency of the i-th ion, F is the Faraday, and c_i is the concentration of the i-th ion in moles per cm^3. It is evident that in the bulk of the solution $\rho = 0$ (condition of electro-neutrality). Besides, the condition of electro-neutrality of the entire system makes it necessary for

* In this and subsequent sections it is assumed for the sake of simplicity that the potential of the wall surface is equal to the potential inside the solid phase, *i.e.* that no diffuse layer exists on the side of the wall.

the surface charge e of the outer phase to have the same absolute value as $\int_{x=0}^{x=\infty} \rho \, \mathrm{d}x$:

$$|e| = \left| \int_{x=0}^{x=\infty} \rho \, \mathrm{d}x \right| . \tag{5.23}$$

The relation between the potential φ at any given point and the space charge density ρ is given by Poisson's equation:

$$\Delta\varphi = -\frac{4\pi}{\varepsilon}\rho . \tag{5.24}$$

Moreover, according to Boltzmann's equation, the concentration of the i-th ion should be

$$c_i = c_{i\infty} \, \mathrm{e}^{-Fz_i\varphi/RT} , \tag{5.25}$$

where $c_{i\infty}$ is the bulk concentration, and $Fz_i\varphi$ is the energy required to transfer 1 mole of the positive ions from the bulk of solution with $\varphi = 0$ to a point with potential φ.

Solution of these equations is simplified in the case of a plane surface, where the Laplace operator Δ is reduced to $\mathrm{d}^2/\mathrm{d}x^2$, allowing (5.24) to be written in the form

$$\frac{\mathrm{d}^2\varphi}{\mathrm{d}x^2} = -\frac{4\pi}{\varepsilon}\rho . \tag{5.26}$$

In accordance with the qualitative prerequisites of the theory explained above, the following boundary conditions exist for the solution of this differential equation:

$$\begin{aligned} x = 0, \ \varphi = \varphi_0, \ \frac{\mathrm{d}\varphi}{\mathrm{d}x} &= \left(\frac{\mathrm{d}\varphi}{\mathrm{d}x}\right)_0 \\ x = \infty, \ \varphi = 0, \ \frac{\mathrm{d}\varphi}{\mathrm{d}x} &= 0 . \end{aligned} \tag{5.27}$$

Let us analyse the case (convenient for calculations) of a binary electrolyte for which $|z_1| = |z_2| = z$. By combining the foregoing equations we obtain

$$\frac{\mathrm{d}^2\varphi}{\mathrm{d}x^2} = -\frac{4\pi}{\varepsilon} Fzc_\infty (\mathrm{e}^{-Fz\varphi/RT} - \mathrm{e}^{+Fz\varphi/RT}) = \frac{8\pi Fz}{\varepsilon} c_\infty \sinh\left(\frac{Fz\varphi}{RT}\right) . \tag{5.28}$$

References p. 171–172

After substituting

$$\frac{d^2\varphi}{dx^2} = \frac{d\left(\frac{d\varphi}{dx}\right)}{dx} = \frac{d\varphi}{dx} \cdot \frac{d}{d\varphi} \cdot \left(\frac{d\varphi}{dx}\right)$$

we obtain:

$$\frac{d\varphi}{dx} d\left(\frac{d\varphi}{dx}\right) = \frac{8\pi RT}{\varepsilon} c_\infty \sinh\left(\frac{Fz\varphi}{RT}\right) d\left(\frac{Fz\varphi}{RT}\right);$$

the first integration* results in the expression

$$\left(\frac{d\varphi}{dx}\right)^2 = \frac{16\pi RT}{\varepsilon} c_\infty \left[\cosh\left(\frac{Fz\varphi}{RT}\right) - 1\right], \tag{5.29}$$

since at $x = \infty$, $\varphi = 0$, $d\varphi/dx = 0$ and $\cosh 0 = 1$. Therefore

$$\frac{d\varphi}{dx} = -\sqrt{\frac{16\pi RT}{\varepsilon} c_\infty \left[\cosh\left(\frac{Fz\varphi}{RT}\right) - 1\right]} =$$

$$-\sqrt{\frac{8\pi RT}{\varepsilon}} c_\infty \left(e^{Fz\varphi/2RT} - e^{-Fz\varphi/2RT}\right) \tag{5.30}$$

since $\cosh x - 1 = \frac{e^x + e^{-x}}{2} - 1 = \frac{e^x + e^{-x} - 2}{2} = \frac{(e^{x/2} - e^{-x/2})^2}{2}$.

The minus sign in front of the square root indicates that φ decreases with increasing distance from the interface.

Some important conclusions may be drawn even from the relation (5.30). In the first place, it is found that the slope $d\varphi/dx$ increases in proportion to $\sqrt{c_\infty}$. The concentration c_∞ obviously applies also to ions which do not affect the potential φ_0. Thus, when the concentration of such ions is increased, the curve $\varphi(x)$ starts again from φ_0 (at $x = 0$); for large c_∞ the curve descends more steeply, and therefore for large c_∞(2) it lies lower than for the more dilute solutions (1) (Fig. 40).

In accordance with the principles of hydrodynamics the velocity of the liquid is zero at the wall, owing to adhesion, and increases in the bulk of the solution.** It is obvious that for the electrokinetic phenomena to take

* Assuming that ε is independent of x.

** This fact is also in opposition to the idea of the molecular electrical double layer as a source of electrokinetic phenomena.

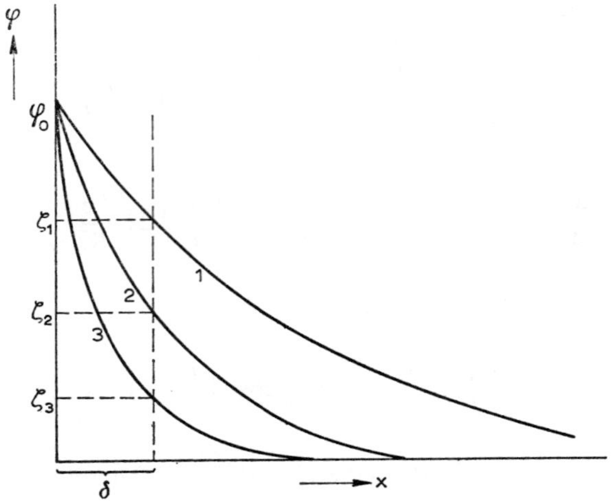

Fig. 40. Distribution diagram of the φ potential at interface according to the Gouy–Chapman theory.

place it is necessary to have a certain drift of the charged liquid with respect to the wall. Let us assume that this drift occurs at a distance δ from the interface (slipping plane), δ being determined by the hydrodynamic conditions of the flow. The electrokinetic potential will then be $\zeta = \varphi_\delta$. It can be seen from Fig. 40 that although the potential φ_0 is not affected by c_∞, the ζ-potential decreases with rising c_∞. In other words, we have not only discovered the essential difference between φ_0 and ζ, but also established the quantitative relationship between ζ and c_∞. These considerations are of course schematic and rather simplified. In any case, the variations of c_∞ influence the potential φ_0, even if only because of the change in the activity of the potential-determining ions, and the above arguments refer therefore only to sufficiently diluted solutions. Furthermore, the magnitude δ has been introduced here without great precision, which is unfortunately as yet unattainable on the basis of hydrodynamics, so that the "quantitative dependence" between ζ and c_∞ should only be regarded as quantitative in the provisional sense.

The above discussion of $d\varphi/dx$ makes it also possible to determine the range of the variation of ζ with electrolyte concentration. At maximum dilutions ζ approaches φ_0 whereas at higher c_∞ it tends to zero, so that $\varphi_0 > \zeta > 0$. It is clear, therefore, that ζ always preserves the sign of the

References p. 171–172

potential φ_0. We have seen (p. 154) that this requirement, which follows from Gouy and Chapman's theory, is not always observed.

The relation derived for $d\varphi/dx$ makes it possible to calculate the surface charge of the opposite phase in the form (5.23). By expressing ρ from the Poisson equation, we obtain

$$e = \int_0^\infty \frac{\varepsilon}{4\pi} \frac{d^2\varphi}{dx^2} dx = \int_0^\infty \frac{\varepsilon}{4\pi} d\left(\frac{d\varphi}{dx}\right) = -\frac{\varepsilon}{4\pi}\left(\frac{d\varphi}{dx}\right)_{x=0}$$

since $(d\varphi/dx)_{x=\infty} = 0$. It should be noticed that e and $\int \rho\, dx$ are of opposite sign. By substituting the value of $d\varphi/dx$ found from (5.30) for $x = 0$ ($\varphi = \varphi_0$), we obtain

$$e = \sqrt{\frac{\varepsilon RT}{2\pi} c_\infty} \left[\exp\left(\frac{zF\varphi_0}{2RT}\right) - \exp\left(-\frac{zF\varphi_0}{2RT}\right)\right]. \tag{5.31}$$

For sufficiently small φ_0, ($zF\varphi_0/2RT \ll 1$), the exponentials may be expanded into a series which, as a first approximation, yields:

$$e = \sqrt{\frac{\varepsilon RT}{2\pi} c_\infty} \frac{zF\varphi_0}{RT} = \frac{\varepsilon}{4\pi}\sqrt{\frac{8\pi z^2F^2c_\infty}{\varepsilon RT}}\varphi_0 = \frac{\varepsilon}{4\pi}\varkappa\varphi_0 \tag{5.32}$$

where

$$\varkappa = \sqrt{\frac{8\pi z^2F^2c_\infty}{\varepsilon RT}}. \tag{5.33}$$

The surface charge is proportional to the surface potential, and the double layer capacity is equivalent to the capacity of a plane condenser with plate distance $1/\varkappa$[*cf.* (5.1)].

Further integration can be carried out as follows. Expressing $d\varphi/dx$ as

$$\frac{d\varphi}{dx} = -\sqrt{\frac{8\pi RT}{\varepsilon} c_\infty} \left[\exp\left(\frac{zF\varphi}{2RT}\right) - \exp\left(-\frac{zF\varphi}{2RT}\right)\right] =$$

$$= -2\sqrt{\frac{8\pi RT}{\varepsilon} c_\infty} \sinh\left(\frac{zF\varphi}{2RT}\right)$$

and separating the variables, we obtain

$$\frac{d\left(\frac{zF\varphi}{2RT}\right)}{\sinh\left(\frac{zF\varphi}{2RT}\right)} = -\frac{zF}{RT}\sqrt{\frac{8\pi RT c_\infty}{\varepsilon}} dx = -\varkappa\, dx.$$

which, after integration $\left(\int \frac{\mathrm{d}x}{\sinh x} = \ln \tanh \frac{x}{2}\right)$ gives:

$$\frac{\tanh\left(\frac{zF\varphi}{4RT}\right)}{\tanh\left(\frac{zF\varphi_0}{4RT}\right)} = \mathrm{e}^{-\varkappa x} \tag{5.34}$$

since $\varphi = \varphi_0$ at $x = 0$.

For small φ_0 (and still smaller φ) both hyperbolic tangents may be expanded into series

$$\left(\tanh x = \frac{\mathrm{e}^x - \mathrm{e}^{-x}}{\mathrm{e}^x + \mathrm{e}^{-x}} \approx \frac{1 + x - 1 + x}{1 + x + 1 - x} = x\right)$$

so that

$$\varphi = \varphi_0 \mathrm{e}^{-\varkappa x}. \tag{5.35}$$

This equation shows that $1/\varkappa$ corresponds to the concept of the "radius of the ionic atmosphere" (in this case to the "thickness of the ionic atmosphere") in the theory of strong electrolytes*.

Eq. (5.34) may also be used to obtain another convenient approximation, which we shall utilise following the example of Overbeek[13]. In the region of $\varphi(x)$ where φ is sufficiently small (so that $zF\varphi/4RT \ll 1$), we obtain by expanding only $\tanh(zF\varphi/4RT)$ into a series,

$$\varphi = \frac{4RT}{zF} \tanh\left(\frac{zF\varphi_0}{4RT}\right) \mathrm{e}^{-\varkappa x}. \tag{5.36}$$

In ref. 13 $\tanh(zF\varphi_0/4RT)$ is denoted as

$$\gamma = \frac{\{\exp(zF\varphi_0/2RT)\} - 1}{\{\exp(zF\varphi_0/2RT)\} + 1}$$

the equivalence of which is shown by

$$\frac{\mathrm{e}^{2x} - 1}{\mathrm{e}^{2x} + 1} = \frac{\mathrm{e}^{-x}(\mathrm{e}^{2x} - 1)}{\mathrm{e}^{-x}(\mathrm{e}^{2x} + 1)} = \frac{\mathrm{e}^x - \mathrm{e}^{-x}}{\mathrm{e}^x + \mathrm{e}^{-x}} = \tanh x.$$

The electrical double layer can also be defined by its electric capacitance.

* It should be noted that the theory of strong electrolytes, which is based on the analogous concept of diffuse structures of ionic clouds forming around every ion, was put forward when Gouy and Chapman's theory was already in existence.

This property is equivalent to the capacitance of a condenser and may be defined by:

$$\text{capacitance of the double layer} = \frac{e}{\varphi_0} \tag{5.37}$$

This magnitude can be measured relatively easily and accurately by methods similar to those used for measuring the capacitance of condensers. Gouy and Chapman's theory allows a calculation of $de/d\varphi_0$ by differentiating the surface charge [*e.g.* as per (5.31)] with respect to the potential of the diffuse layer. $de/d\varphi_0$ can easily be transformed into e/φ_0. In the zero point of charge

$$\frac{e}{\varphi_0} = \frac{\varepsilon\varkappa}{4\pi}. \tag{5.38}$$

At room temperature, and for a concentration of 1 mole of a uni–univalent electrolyte per litre, $\varkappa \approx 3 \cdot 10^7$ [according to (5.33) in which c_∞ is the concentration of the electrolyte in moles/cm^3], resulting in $de/d\varphi_0$ of about 200 μF/cm^2. The experimentally determined capacitances of double layers are of the order of 20 μF/cm^2.

It may thus be stated that although Gouy and Chapman's theory provides a good explanation for the qualitative difference between the φ_0 and ζ potentials, and for the sensitivity of the ζ-potential to the concentration of the non-potential-determining ions, it cannot explain the frequently observed change of the sign of ζ with increasing concentration of some electrolytes. Furthermore, the theory leads to capacitance values of the double layers which in concentrated solutions are too high by one order of magnitude. These divergences between theory and experiment find a good qualitative explanation in the theory of Stern, which is a further development of the double layer theory.

In 1924 Stern[32] pointed out that the ions forming the diffuse double layer are distributed in it not only under the influence of the electric field, but also (especially when very near to the interface) under the influence of the intermolecular forces between them and the outer phase. As demonstrated by adsorption phenomena, the forces of attraction between ions in solution and the outer phase can be quite considerable, and can give rise to the formation of an adsorption layer (usually a monolayer). If such a layer is formed by the adsorption of ions of a certain type, this is equivalent to the accumulation of a certain amount of electricity at the interface, so that the potential near the surface will be changed, as a result of adsorption forces, from φ_0 to a certain value φ_Δ. The potential φ_Δ is determined mainly by adsorption forces and, particularly in the presence of strong adsorption of gegenions* from the outer phase, a "charge reversal of the surface" is possible, *i.e.* reversal of the sign of φ_Δ with respect to the sign of φ_0. Furthermore, taking into account the finite dimensions of ions (which are determined by the forces of repulsion), Stern provided a "geometrical" interpretation of the distribution of φ: at a distance of the order of a molecular diameter from the interface the potential φ_Δ is established, which decreases at further distances in accordance with the theory of Gouy and Chapman. Fig. 41 shows two different courses of $\varphi(x)$: (*i*) when the adsorbed ion is of the same sign as the charge of the outer phase, φ_Δ decreases in relation to φ_0; (*ii*) when a gegenion is adsorbed in a quantity which "reverses the charge of the surface", φ_Δ acquires a sign opposite to φ_0. Intermediate states are possible.

In order to construct a quantitative theory Stern made use of the fact that the adsorption forces decrease very rapidly with increasing distance. We can therefore assume that their role is almost completely negligible for $x > \Delta$ (of the order of a few ionic diameters) and that beyond this distance the ionic distribution is described merely by Gouy and Chapman's theory. In this case the condition of the electrical neutrality of the system is written as:

$$e = -(e_1 + e_2), \tag{5.39}$$

where e_1 is the charge of the adsorption layer, e_2 is the charge of the diffuse part, and e is the charge of the outer phase per cm^2. The surface charge e_2 is calculated as in (5.31), with the difference that integration is carried out not from $x = 0$, but from $x = \Delta$, and correspondingly not from φ_0 but from φ_Δ:

$$e_2 = \sqrt{\frac{\varepsilon RT}{2\pi} c_\infty} \left[\exp\left(\frac{zF\varphi_\Delta}{2RT}\right) - \exp\left(-\frac{zF\varphi_\Delta}{2RT}\right)\right]. \tag{5.40}$$

To calculate e_1 Stern assumed that the adsorption was monomolecular, and made use of Langmuir's equation:

$$e_1 = \frac{ZzF}{1 + \dfrac{1}{c_\infty}\exp\left(\dfrac{zF\varphi_\Delta + \theta_+}{RT}\right)} - \frac{ZzF}{1 + \dfrac{1}{c_\infty}\exp\left(-\dfrac{zF\varphi_\Delta - \theta_-}{RT}\right)}, \tag{5.41}$$

* Gegenions = counterions; that is, ions having a charge opposite in sign to that of the surface.

References p. 171–172

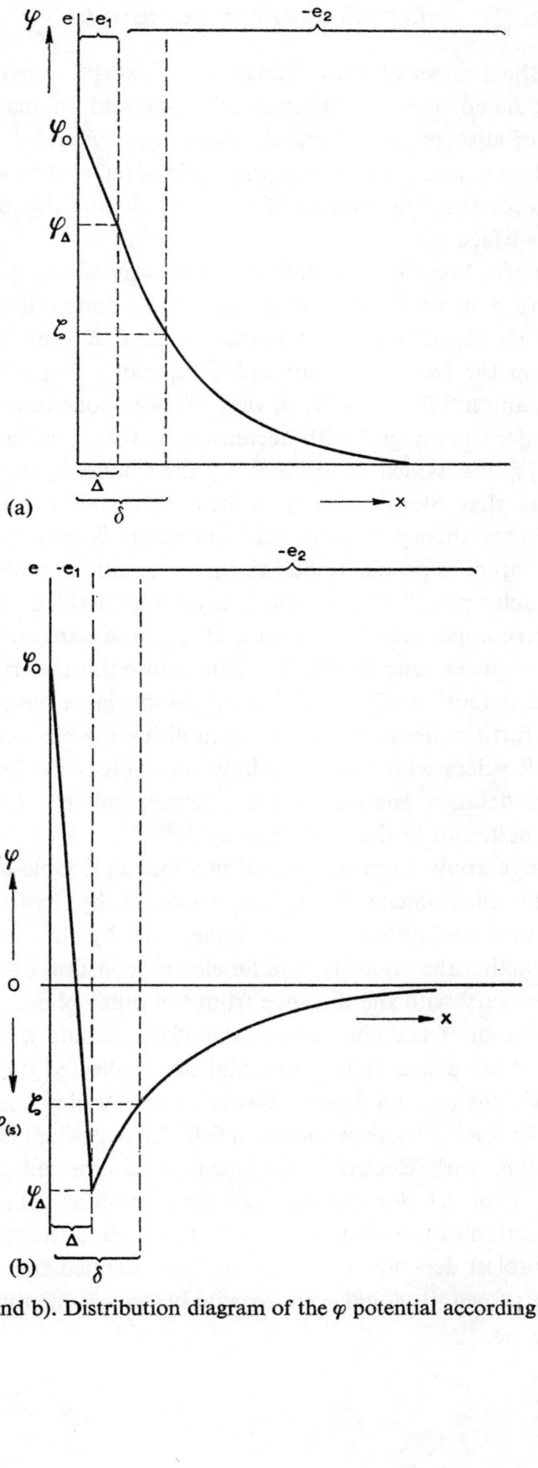

Fig. 41 (a and b). Distribution diagram of the φ potential according to Stern.

where Z is the number of sites available for adsorption per 1 cm^2 and θ_+ and θ_- are the adsorption potentials of cation and anions (equivalent to the energy of adsorption without electric energy $zF\varphi_\Delta$).

Fig. 41 shows clearly how the considerations referred to above allow an explanation for the sign reversal of ζ by introducing the distance δ from the sliding surface.

Furthermore, Stern's correction explains why Gouy and Chapman's theory led to high values for the charge of the double layer, and consequently for its capacitance. This is due to the fact that $|e| = |\int \rho \mathrm{d}x|$ is calculated on the basis of Gouy and Chapman's theory by integrating from $x = 0$, and not from $x = \Delta$. In view of the exponential increase of the function under the integral with decreasing x, this gives large values for the term $\int_{x=0}^{x=\Delta} \rho \mathrm{d}x$, which is replaced by the smaller e_1 in Stern's theory.

It is clear that Stern's theory, which overcame the difficulties encountered in the theory of Gouy and Chapman, is only qualitative. This qualitative nature is primarily due to the schematic manner in which the charge of each "plate" of the double layer was divided into adsorption and diffusion components. Moreover, utilisation of Langmuir's adsorption isotherm introduces considerable limitations into this theory. The problem of the true distribution of potential in the double layer has not as yet been solved. No further discussion of this complicated subject can be given in this book. Readers who wish to follow these ideas further are referred to the more detailed analysis of the electrochemistry of double layers, which may be found in the bibliography[13,33,34].

Quantitative application of the still incomplete double layer theory to electrokinetic phenomena is further hindered by hydrodynamic uncertainties with respect to the slip plane, and by the fact that we do not know whether the viscosity η and dielectric constant ε are constant, or whether they vary with the distance from the outer phase.

The present brief account cannot therefore include a review and an assessment of the available experimental data collected during studies of elektrokinetic phenomena. Even in Kruyt's very detailed book[13] Overbeek was unable to resolve these problems in full. An overwhelming majority of studies dealing with electrokinetic phenomena are reduced to an investigation of the relationship between the calculated zeta potential and the concentration of the electrolyte. Oddly enough, although this appears to be the simplest dependence, it has not been defined theoretically in an unambiguous form. It would seem reasonable to assume that the distance

References p. 171–172

δ between the slip plane and the outer phase, determined by hydrodynamic factors, should be identical for all dilute aqueous solutions and a smooth solid surface. In actual fact, Eversole and Boardman's attempts (1941)[35] to calculate δ from the data obtained during a study of ζ (c) relationships yielded an inexplicable result: δ was found to vary between 8 and 100 Å.

In spite of these difficulties, investigations of electrokinetic phenomena possess considerable inherent interest, *e.g.* for the phenomena of transfer in porous media, which are highly important in practice. Some ideas concerning this vast field can be obtained, *e.g.* from ref. 36.

The above brief account of the structure of electrical double layers is sufficient for an understanding of some essential peculiarities in the behaviour of thin layers, and particularly, for the explanation of the phenomena occurring during the slow coagulation of lyophobic colloids. These topics will be treated later in the book.

3. Measurement of Electrophoretic Mobilities

It is preferable to investigate electrokinetic phenomena on the basis of electro-osmosis and of the streaming potential, since these two phenomena possess a better theoretical foundation. However, in view of the more special significance of such investigations, no attempt will be made to analyse them in the present work. A review may be found in the already quoted book by Kruyt[13].

An account will therefore now be given of the theoretically less clear, though nevertheless widely applied methods by which electrophoresis may be studied. A fuller description of such methods is again given by Kruyt[13].

The electrophoresis of colloidal particles is very similar to ionic migration in an electric field, not only with respect to the nature and mechanism of the phenomenon, but also as regards the magnitude of electrophoretic mobility which, similarly to ionic mobility, is of the order of microns per second when the electric field strength is 1 V/cm. This points also, if not to the independence of, then certainly to a weak dependence of electrophoretic mobility on the particle size.

In spite of the similarity between electrophoresis and ionic migration, the methods adopted so far for the investigation of these phenomena are rather different. The moving boundary method, seldom used for ionic migration, was found to be extremely fruitful for electrophoresis. In a number of cases, electrophoresis can be studied directly by microscopic or ultramicroscopic methods; this is impossible in the case of ionic migration, in view of the submicroscopic dimensions of the ions.

The possibility of applying the moving boundary method, in which the rate of electrophoretic migration is determined from measurements of the displacement of the boundary between the colloidal solution and its ultrafiltrate in the presence of an electric field, is due to the fact that the conductance of the colloidal solution is usually only slightly greater than that of the pure dispersion medium. This follows naturally from the fact that, for mobility of colloidal particles which is roughly the same as the ionic mobility of ions, in a given electric field, the colloidal particles (owing to their larger dimensions) give rise to much smaller particle concentrations than the ions and therefore participate to a relatively lesser extent in the overall transfer of electricity through the solution. This property is highly important, since if the two solutions forming the boundary, whose velocity is taken as a measure of the mobility of a given component in an electric field, exhibit different conductances, then the electric field changes abruptly at the boundary, and also varies with time as the boundary is displaced. Although such non-homogeneity and dependence of the field strength on time hinders considerably the application of moving boundary methods to ionic migration, the difficulty either does not arise at all or only arises to a minor extent during electrophoresis. Application of the moving boundary method to colloidal solutions possesses important advantages, allowing not only a determination of the electrophoretic mobility, but also a separation of the components endowed with different mobilities during the process of electrophoresis. The nature and quantity of each component may also be determined. As a result, the moving boundary method has been carefully developed by Tiselius in 1930[37], and electrophoresis on paper and in other non-liquid media has found wide application.

The U-shaped tube of the Tiselius apparatus, shown in Fig. 42 in section and plan views, consists of several sections which can slide relative to each other in the horizontal direction along planes aa′, bb′, and cc′. The channel forming the U-shaped cavity (thick line), has a rectangular cross section. To obtain a boundary between the colloidal solution and the ultrafiltrate, the solution is first poured into B, up to the level bb′. Section C is moved sideways, washed, filled with ultrafiltrate, and is then returned to the initial position by means of a special appliance which ensures

References p. 171–172

smooth movement. A sharp solution/ultrafiltrate boundary is thus obtained. Before matching with B, C is connected to the electrode compartments with Ag/AgCl electrodes immersed in a concentrated solution of an alkali metal chloride. After the boundary has been formed, the source of emf is connected to the electrodes.

The apparatus described above ensures not only a distinct boundary at the beginning of the experiment, but also the possibility of using relatively high field values, *i.e.* of carrying out the electrophoresis at a fast rate conducive to preserving the sharpness of the boundary during its displacement. To achieve this end, the electrode compartments and electrodes themselves are large, and are removed from the space in which the electrophoresis takes place, in order to prevent the products of electrolysis from influencing the state in the U-shaped part of the apparatus. To reduce thermal convection which would diffuse the boundary, the operation is carried out at a temperature close to 4° C (maximum density of aqueous

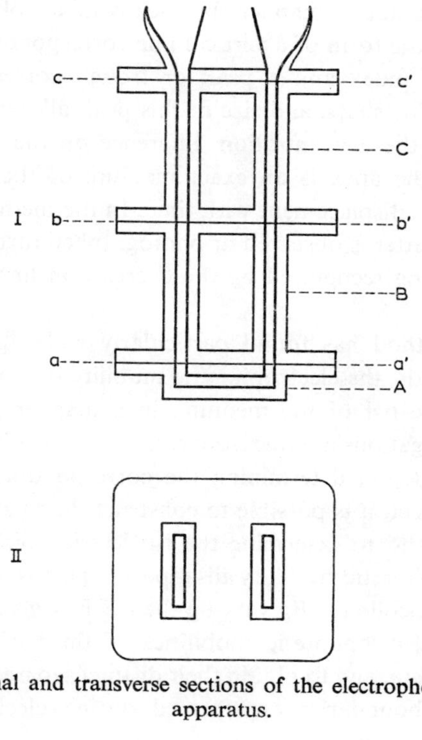

Fig. 42. Longitudinal and transverse sections of the electrophoretic cell in Tiselius' apparatus.

solutions), where slight temperature differences have only a slight effect on the density. The rectangular cross-section of the U-shaped channel allows an accurate optical check of the distribution of concentration during electrophoresis if the walls are of good optical quality.

Two methods are used to observe the movement of the boundary and the concentration distribution, both of which are based on the variation of the refractive index of the solution with concentration. These are Tepler's "shadow" method, *e.g.* in the Philpot[38]–Svensson[39] variant (1938–1939), and Lamm's scale method (1937)[40]. These methods are based on the fact that when a parallel beam of light traverses a solution in the boundary zone, where a gradient of concentration and therefore of refractive index has been established, the rays are bent in the direction of the higher refractive index. In consequence, when a beam of light is passed through a cell containing the solution, a screen placed behind the cell will show not only the basic image of the source (a horizontal line), but also a certain diffuseness (downwards or upwards). The latter can be effectively recorded in a quantitative manner, by means of an oblique slit and a cylindrical lens, in the form of a vertical line corresponding to regions of constant refractive index and a peak or trough corresponding to the boundary region. The shape and size of this peak allow an estimation of the sharpness and the concentration difference on the two sides of the boundary, whilst the apex is an exact measure of the position of the boundary and of its displacement with time. In the method involving the use of a scale, the latter is observed or photographed through the cell, the boundary zone being recognised by the decrease in line spacings of the scale.

The Tiselius method has found particularly wide application in the studies of biocolloids, the electrophoretic mobility of which depends to a large extent on the pH of the medium, in a manner specific for each colloid. The investigations are therefore carried out in a buffered medium. By varying the pH and determining the corresponding electrophoretic mobility of the object, it is possible to construct the relationship between mobility and pH and to determine the isoelectric point, *i.e.* the pH at which the electrophoretic mobility disappears. This is a very important constant for the biocolloids. By varying the pH it is also possible to vary the ratio of the electrophoretic mobilities of the various components present in the mixture, and to obtain their distinct separation owing to the fact that several boundaries are formed during electrophoresis, cor-

References p. 171–172

responding to the number of components with sufficiently different u_e.

The wide possibilities of macro-electrophoresis described above gave rise to one more method, paper electrophoresis, which although less exact is considerably more simple. The investigated solution of a biocolloid is applied as a vertical line or a spot on to a strip of filter paper moistened with the buffer solution. The strip is placed horizontally in an enclosed space, and its ends are immersed in vessels containing the buffer and electrodes. When the source of emf is connected to the electrodes, the field generated along the strip causes the components in the line or spot to be displaced in accordance with their electrophoretic mobility. After some time the electrophoresis is stopped, the strip is dried, and is then immersed in a solution of a dye which is adsorbed preferentially on the biocolloid. The resulting image of the positions of the various components at the end of electrophoresis may be used to assess the number of components and their electrophoretic mobilities. It can be seen that the paper plays the role of a porous medium which prevents the spreading of components and their convective mixing with the medium in which electrophoresis takes place. An important advantage of paper electrophoresis over the moving boundary method consists in that the components can be completely separated by elution of the corresponding zones. Continuous separation methods have been evolved by combining the spread along the inclined filter paper with electrophoretic movement. Latterly, paper has been replaced by some gelated media (agar-agar, gelatin), which yield more distinct zones. Electrophoresis on paper and in other media is accompanied by side phenomena, such as the transfer caused by migration of the evaporating buffer [Macheboeuf, Rebeyrotte, Dubert (1953)[41]]. It was also established [Sheludko, Konstantinov, Tsvetanov (1959)[42]] that, *e.g.* in gelatin, the electrophoretic mobility of some dyes is not only lower than in water or aqueous solutions, but that the mobility ratios of the components are entirely different. These peculiarities have not as yet been studied in great detail. Problems of the theory of electrophoresis in porous and gelated media, which is now in the process of development, and the diverse experimental techniques are currently studied in view of the practical importance of the method. Some idea of these investigations can be found *e.g.* in the monograph by Lederer[43].

In cases when the colloidal particles are visible under a microscope or an ultramicroscope, it is possible to determine their electrophoretic mobility by observing the movement of an individual particle in the electric field.

This direct method cannot unfortunately be applied to the most important biocolloids owing to the small sizes of the particles concerned. In view of the very short focal distance of the objective at large magnifications, microscopic observation of particles must necessarily involve working with very narrow plane-parallel or cylindrical cells. In such cells the applied electric field gives rise not only to the electrophoresis of particles, but also to electro-osmotic transfer of the liquid, so that the observed overall displacement of particles is the sum of the electrophoretic and electro-osmotic movement. If the cell is of the closed type, electro-osmotic flow along the walls should be compensated by the reverse hydrodynamic flow within the cell. It is obvious that in this case there will exist levels at which velocity of the liquid flow is equal to zero symmetrically to the plane of symmetry if the cell has the form of a parallelepiped. If this level is known and if the electrophoretic mobility in it is measured, one can obtain directly the purely electrophoretic velocities of particle displacement. This problem was solved by Smoluchowski[10]. According to eqn. 3.35, with the appropriate integration constant, the velocity of flow at a given level $x < r$ in a cylindrical cell is

$$u = u_{eo} - c(r^2 - x^2) \tag{5.42}$$

where u_{eo} is the electro-osmotic velocity, $c(r^2 - x^2)$ is the velocity of the reverse flow according to Poiseuille's law and r is the radius of the cell. The constant c is determined from the condition that the total flow must be equal to zero:

$$\int_0^r 2\pi x u \, dx = 0, \quad i.e. \quad c = 2u_{eo}/r^2,$$

hence

$$u = u_{eo}\left[2\left(\frac{x}{r}\right)^2 - 1\right]. \tag{5.43}$$

Thus at the walls ($x = r$) the velocity of flow is u_{eo}, at the centre ($x = 0$) it is $-u_{eo}$ and at $x = r/\sqrt{2} = 0.354(2r)$ the velocity of flow is equal to zero. If electrophoresis at this level is observed, the obtained values of particle velocity yield the electrophoretic mobility directly when divided by the field strength. Analogous calculations for a sufficiently wide plane-parallel cell yield the zero velocity level for the liquid as $h = 0.212d$, where d is the thickness of the cell. It follows that by carrying out measurements at several (at least two) levels, both electro-osmotic and electrophoretic velocities may be calculated.

References p. 171–172

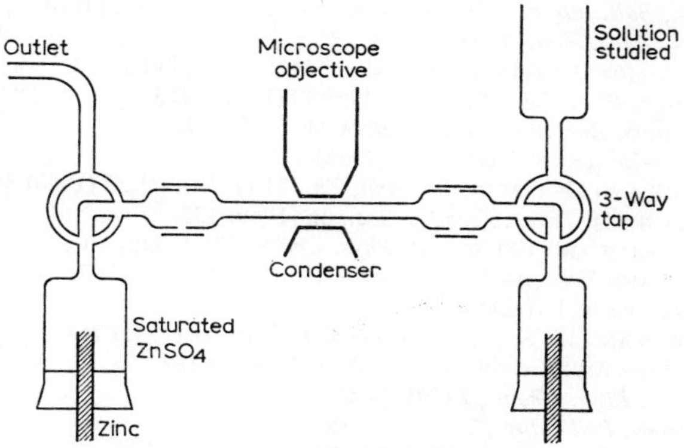

Fig. 43. Micro-electrophoresis apparatus.

A micro-electrophoresis apparatus designed by Abramson in 1929[44] for working with a source of constant emf is illustrated in Fig. 43. Application of alternating electric fields [Svedberg (1919)[45]] also provides numerous possibilities. In this case electrode polarisation and electrolysis are considerably reduced, and the apparatus may thus be considerably simplified. Particle displacements are in such cases expressed as their vibration about their equilibrium positions, projected as lines whose length is proportional to the velocity of particle displacements at various levels in the cell. Unfortunately, this variant of micro-electrophoresis has not been widely used.

REFERENCES

1 Reuss, *Mem. Soc. Natur Mosc.*, 2 (1809) 327.
2 Wiedemann, *Pogg. Ann. Physik*, 87 (1852) 321.
3 Quincke, *Pogg. Ann. Physik*, 113 (1861) 513.
4 Quincke, *Pogg. Ann. Physik*, 107 (1859) 21.
5 Dorn, *Wied. Ann.*, 10 (1880) 46.
6 Wiedemann, *Pogg. Ann.*, 99 (1856) 177.
7 Hepburn, *Proc. Phys. Soc. (London)*, 39 (1927) 99.
8 Helmholtz, *Wied. Ann.*, 7 (1879) 337.
9 Perrin, *J. Chim. Phys.*, 2 (1904) 601.
10 Smoluchowski, in Graetz (Herausgeber), *Handbuch der Elektrizität und des Magnetismus*, Leipzig, 1914, Vol. 2.
11 Bikerman, *Kolloid-Z.*, 72 (1935) 100.
12 Urban, White and Strassner, *J. Phys. Chem.*, 39 (1935) 311.
13 Overbeek, in Kruyt (Editor), *Colloid Science*, Elsevier, Amsterdam, 1952.
14 Freundlich and Abramson, *Z. Phys. Chem.*, 128 (1927) 25; 133 (1928) 51.
15 Overbeek (Editor), *Advan. Colloid Sci.*, Vol. 3 (1950).
16 Mooney, *Phys. Rev.*, 23 (1924) 396.

17 Stock, *Bull. Intern. Acad. Sci. Cracovie*, A (1913) 131; (1914) 95.
18 Saxen, *Wied. Ann.*, 47 (1892) 46.
19 Mazur and Overbeek, *Rec. trav. Chim.*, 70 (1951) 83.
20 Onsager, *Phys. Rev.*, 37 (1931) 405; 38 (1931) 2265.
21 Sheludko, *Annuaire Univ. Sofia*, 50 (1955/56) 99.
22 Krasny-Ergen, *Kolloid-Z.*, 74 (1936) 172.
23 Dukhin and Deryagin, *Kolloidn. Zh.*, 21 (1959) 37; 22 (1960) 587.
24 Guggenheim, *Trans. Faraday Soc.*, 36 (1940) 139.
25 Freundlich and Ettisch, *Z. Phys. Chem.*, 116 (1925) 401.
26 Kruyt and Van der Willigen, *Kolloid-Z.*, 45 (1928) 307.
27 Wijga, Thesis, Utrecht, 1946.
28 Rutgers and De Smet, *Trans. Faraday Soc.*, 41 (1945) 758.
29 Troelstra and Kruyt, *Kolloid-Z.*, 101 (1942) 182.
30 Gouy, *J. Phys. Chem.*, 9 (1910) 457.
31 Chapman, *Phil. Mag.*, 25 (1913) 475.
32 Stern, *Z. Elektrochem.*, 30 (1924) 508.
33 Frumkin, Bagotskii, Iofa and Kabanov, *Kinetika elektrodnykh protessov* (*Kinetics of electrode processes*), University of Moscow, 1952.
34 Bockris (Editor), *Modern aspects of electrochemistry*, Butterworth, London, 1954, Vol. 1.
35 Eversole and Boardman, *J. Chem. Phys.*, 9 (1941) 789.
36 *Elektrokineticheskie svoistva kapillyarnykh sistem, sbornik* (*Electrokinetic properties of capillary systems, collection of papers*) Academy of Sciences, USSR, Moscow 1956.
37 Tiselius, Thesis, Uppsala, 1930.
38 Philpot, *Nature*, 141 (1938) 283.
39 Svensson, *Kolloid-Z.*, 87 (1939) 181.
40 Lamm, *Nova Acta Regiae Soc. Sci. Upsaliensis*, 10 (1937) no. 6.
41 Macheboeuf, Rebeyrotte, Dubert and Brunerie, *Bull. Soc. Chim. Biol.*, 35 (1953) 334.
42 Sheludko, Konstantinov and Tsvetanov, *Kolloidn. Zh.*, 21 (1959) 747.
43 Lederer, *Introduction to paper electrophoresis and related methods*, Elsevier, Amsterdam, 1955.
44 Abramson, *J. General Physiol.*, 12 (1929) 469.
45 Svedberg and Andersson, *Kolloid-Z.*, 24 (1919) 156.

CHAPTER VI

Thin Layers

It has already been pointed out in Chapter IV that thin layers provide the simplest and most developed example of small phases with special thermodynamic properties. These layers have a special significance in the theory of the stability of lyophobic colloids, being the most important structural element of concentrated foams and emulsions and, apparently, the determining factor in the process of the adhesion of bubbles to particles suspended in a liquid, which takes place during flotation. Thin layers on a base are produced during polymolecular adsorption.

The special thermodynamic properties of a thin layer of thickness h can be characterised unambiguously by its chemical potential μ_h, which differs from the chemical potential μ_∞ of the same substance in bulk. The excess $\mu_h - \mu_\infty = \Delta\mu_h$ is the fundamental subject of any investigation into the peculiarities of thin layers. The relationship between the chemical potential of the thin layer and other parameters which determine its state can be found as follows.

Variation of the free energy of a thin one-component layer can be written in the usual form (Chapter IV):

$$\mathrm{d}F = -S\,\mathrm{d}T - \mathrm{d}A + \mu\,\mathrm{d}n\,.$$

The mechanical work dA which can be carried out by the layer, which is simultaneously a bulk and a surface phase, consists of two components:

$$\mathrm{d}A = P\,\mathrm{d}V + \Delta\,\mathrm{d}O\,.$$

The first term, $P\mathrm{d}V$, is the work of the volume expansion of a layer with constant mass (n = const) and constant area O. The second term, $\Delta\mathrm{d}O$, is the work of spreading of a layer having a constant volume and a constant mass. This work is wholly analogous to the work of a monolayer which spreads on the surface of the base liquid to shift the barrier in the Langmuir–Adam balance. The quantity Δ is thus the two-dimensional surface pressure which causes the layer to spread. The difference between this case and the case of a bulk phase (Chapter IV, Sec. 1) is that now layers are considered that are so thin that the whole layer is within the transitional interphase region. Consequently, S, P and Δ (which replaces $-\sigma$ in (4.9)) depend on the layer thickness h. There is no reason to distinguish between surface and bulk properties, therefore

$$\mathrm{d}F = -S\,\mathrm{d}T - P\,\mathrm{d}V - \Delta\,\mathrm{d}O + \mu\,\mathrm{d}n$$

and

$$\mu = \left(\frac{\partial F}{\partial n}\right)_{T,V,O}.$$

This expression is further simplified by the fact that the compressibility of liquids is very slight, so that $P\mathrm{d}V$ may be taken as zero without affecting the accuracy. Consequently

$$\mathrm{d}F = -S\,\mathrm{d}T - \Delta\,\mathrm{d}O + \mu\,\mathrm{d}n\,,$$

and the required chemical potential is

$$\mu = \left(\frac{\partial F}{\partial n}\right)_{T,O} \tag{6.1}$$

i.e. the derivative of free energy of the layer with respect to the number of molecules which constitute the layer, at constant temperature T and area O.

The occurrence of an excess chemical potential in the layer $\Delta\mu(h)$, implies also a change of the film pressure (4.12), *i.e.* disregarding compressibility,

$$\Delta\mu(h) = -\Pi\bar{v} \tag{6.2}$$

where Π is the excess pressure in the layer (as compared with an infinite mass), and $\bar{v} = Oh/n$ is the volume per molecule of the substance. It has already been noted that this additional pressure was introduced by Deryagin, who called it the disjoining pressure, as a measure of the change of chemical potential in a thin layer. The phenomenon of "disjoining", *i.e.* the tendency towards the thickening of the layer, arises when the pressure in the layer is lower than that in the infinite phase connected with it. Π and $\Delta\mu$ are therefore of opposite sign*. Making use of the fact that $n = Oh/\bar{v}$ we obtain from (6.1) and (6.2)

* In agreement with Deryagin[1] we propose to denote the disjoining pressure by Π instead of P used earlier by Deryagin, and π used earlier by the present author.

References p. 207

$$\frac{\partial \Pi}{\partial h} = -\frac{1}{O}\left(\frac{\partial^2 F}{\partial h^2}\right)_{O,T} = -\left(\frac{\partial^2 f}{\partial h^2}\right)_{O,T} \tag{6.3}$$

where f is the free surface energy per cm^2.

The change of $\Delta\mu$ of the thin layer results also in the variation of the equilibrium pressure of its vapour p_h as compared with the equilibrium vapour pressure of the infinite phase, p_∞, according to (4.17) for an ideal gas

$$\Delta\mu = RT \ln \frac{p_h}{p_\infty} \tag{6.4}$$

or, in the case of equilibrium with an ideal solution

$$\Delta\mu = RT \ln \frac{c_h}{c_\infty} \tag{6.5}$$

where c_h and c_∞ are the respective concentrations of solutions in equilibrium with the layer of thickness h and with the infinite phase.

Since $\Delta\mu(h)$ disappears in sufficiently thick layers ($h \to \infty$, $\mu_h \to \mu_\infty$), Π also tends to zero, $p_h \to p_\infty$ and $c_h \to c_\infty$.

If a thin layer is applied to the surface of a liquid, then the free surface energy undergoes a change, attaining a certain value $\sigma_h \neq \sigma_0$, where σ_0 is the free surface energy of the liquid base. The corresponding change of surface tension $\sigma_0 - \sigma_h = \Delta$ is in fact a two-dimensional surface pressure, analogous to the surface pressure analysed in detail in connection with monomolecular films of insoluble substances on the surfaces of liquids. The change of free energy associated with the increase of the layer area by dO and with the corresponding decrease of free surface by the same magnitude (the temperature and the number of molecules in the layer remaining constant) is:

$$\left(\frac{\partial F}{\partial O}\right)_{n,T} = -\Delta \tag{6.6}$$

O and n being related by the condition $Oh/\bar{v} = n$.

Since the order of differentiation is immaterial, we obtain from (6.1) and (6.6), at $T = \text{const.}$,

$$\left[\frac{\partial}{\partial O}\left(\frac{\partial F}{\partial n}\right)_O\right]_n = \left[\frac{\partial}{\partial n}\left(\frac{\partial F}{\partial O}\right)_n\right]_O,$$

or

$$\frac{\mathrm{d}\mu}{\mathrm{d}h}\left(\frac{\partial h}{\partial O}\right)_n = -\frac{\mathrm{d}\Delta}{\mathrm{d}h}\left(\frac{\partial h}{\partial n}\right)_O{}^{\dagger}$$

which, after substitution

$$\left(\frac{\partial h}{\partial O}\right)_n = -h/O, \quad \text{and} \quad \left(\frac{\partial h}{\partial n}\right)_O = \bar{v}/O,$$

reduces to

$$\frac{\mathrm{d}\Delta}{\mathrm{d}\mu} = \frac{h}{\bar{v}}{}^{\dagger\dagger} \tag{6.7}$$

The maximum variation of free surface energy at $h \to \infty$, as compared with the free surface energy of a liquid in the absence of a layer, is thus

$$\Delta_\infty = \sigma_{10} - \sigma_{12} - \sigma_{20}, \tag{6.8}$$

where 1 denotes the base, 2 the layer, and 0 the gaseous phase (in this notation $\sigma_0 = \sigma_{10}$).

Apart from the above-mentioned changes of parameters which describe the equilibrium of a thin polymolecular layer and express the change of its chemical potential, changes of other parameters are also possible. Thus for example, the electrochemical potential and the energy required for the separation of an electron should be different for thin and thick metallic layers.

1. Molecular Theory of Thin Layers

The free energy of a thin layer can be written as

$$F = U - TS.$$

If we assume that U is a temperature-independent function of the molecular interactions in the layer and between the molecules in the layer and in the adjacent phases, according to (6.1)

† Partial differentials for μ and Δ have been replaced here by complete differentials since μ and Δ are only functions of h.

†† This equation was suggested by Frumkin[2] (1938) in the form $\mathrm{d}\sigma/\mathrm{d}\Pi = h$ [*cf.* eqn. (6.2)].

References p. 207

$$\mu_h = \left(\frac{\partial F}{\partial n}\right)_{O,T} = \left(\frac{\partial U}{\partial n}\right)_O - T\left(\frac{\partial S}{\partial n}\right)_{O,T} \qquad (6.9)$$

consists of two parts: the temperature-independent term $(\partial U/\partial n)_O$ and the entropic term $T(\partial S/\partial n)_{O,T}$ related to thermal motion. The latter term can be disregarded at relatively low temperatures.

The first term of μ_h can be calculated fairly readily by introducing certain simplifying assumptions and making use of equations describing the interaction between two molecules as a function of their distance apart. The principal simplifying assumption is that the molecular interaction is an independent characteristic, *i.e.* that the total interaction is an additive magnitude. Furthermore, in some calculations the possibility of an ordered crystal-like arrangement of molecules is disregarded, and the layer is analysed as an ideally liquid formation.

Let us first analyse the simplest case of a free liquid film (*e.g.* a foam film) in equilibrium with saturated vapours*. In this case the interaction between film molecules and the gaseous phase may be neglected, since the density of vapours in the gas phase is relatively small, and the interaction energy between gas molecules and the film is insignificant in view of the generally very rapid reduction of intermolecular forces with increasing distance.

The quantity μ_h can be calculated in the usual way [Hamaker for the interaction between two spherical particles (1937)[3,4]], by calculating the interaction energy U and differentiating it with respect to the number of molecules (n) present in the layer. It is however much simpler to calculate the change of free energy occurring when one molecule is removed from the layer, which yields $(\partial U/\partial n)_O$ directly. This change is equal to the bonding energy u of one molecule in the surface layer plus the work of regrouping of its closest neighbours** if the layer is liquid, or to the bonding energy of one molecule in the "mid-crystal" position. However, we are interested not in μ_h but in $\Delta\mu_h = \mu_h - \mu_\infty$, which considerably simplifies the problem. According to what has been said before, the "static", or temperature-independent part is

$$\Delta\mu_h = \left(\frac{\partial U}{\partial n}\right)_O - \left(\frac{\partial U_\infty}{\partial n}\right)_O = u_h - u_\infty .$$

* Just as before and in later sections, the layer is assumed to be thin and of infinite extent.

** Filling of the gap which remains after a molecule has been removed from the surface.

Assuming complete additivity of bonding energies, the difference between the bonding energies per molecule $u_h - u_\infty = -(u_\infty - u_h)$, may be regarded as a difference between the bonding energy of a surface molecule with a semi-infinite phase (u_∞) and its bonding energy with a layer having thickness h, this difference being taken with a minus sign. This is of course nothing else than the energy of the interaction between molecules in a semi-infinite phase and molecules separated from it by a distance h. This leads to two conclusions. In the first place, since $\Delta\mu$ is calculated as the difference $-(u_\infty - u_h)$, the work of regrouping of the nearest neighbours after the separation of a molecule from the liquid film and from the surface of a semi-infinite liquid is eliminated from the calculation, being equal in both cases. Secondly, the possibility of reducing the problem to a calculation of the energy of interaction between the molecules in a semi-infinite phase and a molecule removed to a distance h from this phase, shows that $\Delta\mu$ does not depend on interactions of the closest neighbours, and that there is therefore no reason why integration should not be introduced instead of the summation of individual interactions*. In view of the discrete structure of matter, integration would not be exact if the calculation had to be extended down to the closest neighbours.

Determination of the energy of interaction between a molecule and a semi-infinite phase may be solved by the following simple method (Fig. 44). All molecules within the volume enclosed between the spherical surfaces of radii r and $r + \mathrm{d}r$ are at the same distance r from molecule 1 which is at a distance h from a semi-infinite phase of the same substance. The area of the segment with radius r is $2\pi r(r-h)$, and the

* For sufficiently thick layers.

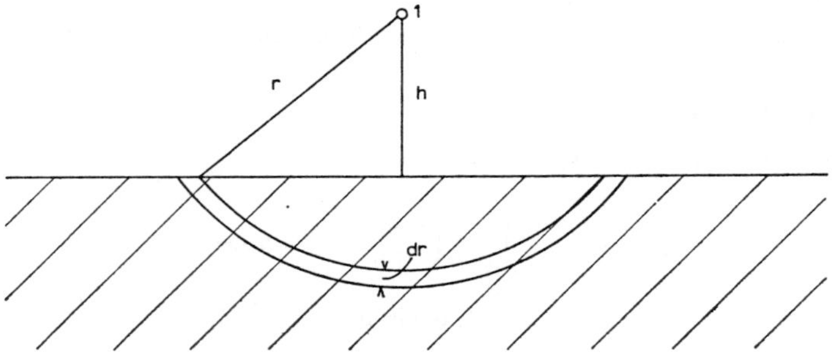

Fig. 44. Determination of the energy of interaction between a molecule and a semi-infinite phase.

References p. 207

volume of the layer of equidistant molecules is therefore $2\pi r(r-h)\mathrm{d}r$. If the number of type 1 molecules in 1 cm^3 of this phase is $1/\bar{v}_1$ and if the energy of interaction between two molecules separated by a distance r is $-\beta_{11}/r^6$, then the total energy of interaction between molecule 1 and all molecules of the same kind in the volume referred to above is

$$-\frac{2\pi\beta_{11}}{\bar{v}_1}\frac{r-h}{r^5}\mathrm{d}r.$$

The energy of interaction with the entire semi-infinite phase is then obtained by integrating the above expression from $r=h$ to $r=\infty$. In accordance with the preceding considerations this yields the required $\Delta\mu$ (with the sign reversed) *i.e.*

$$\Delta\mu = \frac{\pi\beta_{11}}{6\bar{v}_1}\frac{1}{h^3} = \frac{A_{11}\bar{v}_1}{6\pi}\frac{1}{h^3} = \frac{K_{11}\bar{v}_1}{h^3}, \tag{6.10}$$

where

$$A_{11} = \frac{\pi^2\beta_{11}}{\bar{v}_1^2} \quad \text{and} \quad K_{11} = \frac{\pi\beta_{11}}{6\bar{v}_1^2}.^\dagger$$

This result can be easily extended to the case of a thin layer 1 lying on the surface of a semi-infinite phase 2. In this case the bonding energy of a molecule in the surface layer is increased by the bonding energy of this molecule 1 with the semi-infinite phase 2 at distance h away. The absolute value of this increase obviously agrees with (6.10) with $\bar{v}=\bar{v}_2$ for phase 2 and at $\beta=\beta_{12}$ for the constant of the interaction between molecule 1 and molecule 2, *i.e.*

$$-\frac{\pi\beta_{12}}{6\bar{v}_2}\frac{1}{h^3}.$$

The sum which represents $\Delta\mu$ of a layer supported on a substrate is therefore

$$\Delta\mu = \left(\frac{\pi\beta_{11}}{6\bar{v}_1} - \frac{\pi\beta_{12}}{6\bar{v}_2}\right)\frac{1}{h^3} = \frac{(K_{11}-K_{12})\bar{v}_1}{h^3}. \tag{6.11}$$

where $K_{12}=\pi\beta_{12}/6\bar{v}_1\bar{v}_2$.

† The constant A_{11} was introduced by Hamaker, whereas the constant K_{11} was suggested by the present author, as the proportionality coefficient in the relation $\Pi = -K_{11}/h^3$, and was called the "van der Waals' constant".

In the first case, (6.10), the sign of $\Delta\mu$ is fixed, *i.e.* $\mu_h > \mu_\infty$ and correspondingly $\Pi < 0$ and $p_h > p_\infty$ in accordance with (6.2) and (6.4).

In the second case, (6.11), the sign is indefinite, and μ_h can be either greater or smaller than μ_∞. If the base 2 is indistinguishable from the layer 1, *i.e.* if $\beta_{11}/\bar{v}_1 = \beta_{12}/\bar{v}_2$, then obviously $\Delta\mu = 0$.

The case of the free layer 1 (6.10) is equivalent to a plane-parallel gap, of the same thickness h, between two semi-infinite phases 1. In the latter case the bonding energy of a molecule on the surface of one of the phases is equal to its bonding energy with the semi-infinite phase minus* the bonding energy of the same molecule with another semi-infinite phase a distance h away. If we deduct from this (*i.e.* from μ_h) the bonding energy with the semi-infinite phase on which the molecule is resting (μ_∞)**, then the remainder, which is equal to the energy of interaction between the semi-infinite phase and a molecule at a distance h (with the minus sign) is equal to $\Delta\mu$ of the layer with a thickness h.

In all previous calculations it was assumed that the energy of interaction between any two molecules varied inversely with the sixth power of the distance, in accordance with London's law of intermolecular attraction. These calculations can be easily extended to the more general case in which the interaction varies inversely with the m-th power of the distance when $m > 3$. For $m \leqq 3$ the total energy of interaction is found to be infinite, *e.g.* in the case of Coulomb forces ($m = 1$), and the problem becomes very complex. For molecular forces m is always very large ($m > 3$) and such complications do not arise. A full calculation of interactions also presupposes a consideration of the forces of repulsion, which decrease even more rapidly with increasing distance, *e.g.* as the ninth or even higher power. The expression $(\alpha_{ik}/r^9 - \beta_{ik}/r^6)$ should be used in such cases in place of $-\beta_{ik}/r^6$. It is clear that when h is sufficiently large in relation to intermolecular distances the contribution due to the repulsion forces becomes negligible, in view of their much steeper decrease with increasing distance. If on the other hand the calculations are continued down to *e.g.* monomolecular layer thicknesses, the repulsion must be taken into account. In this case it is necessary to determine the minimum intermolecular distance r_0, which can be obtained either from the condi-

* The subtraction is due to the fact that the attraction forces exerted by the two phases act in opposite directions.

** Since this is equivalent to removing the second phase to an infinite distance, *i.e.* to an infinitely wide gap.

tion $d(\alpha_{ik}/r^9 - \beta_{ik}/r^6)/dr = 0$* or from the density. Such calculations are in conflict with other more complex factors, such as the dubious validity (as had been noted before) of substituting integration for summation, and the necessity of an allowance for the molecular kinetic motion, *i.e.* the entropy component which increases in very thin layers and changes its nature on transition to the monolayer. These highly complex problems have not yet been solved and will not be discussed.

Calculations of the "static", temperature-independent part of $\Delta\mu$ of sufficiently thick layers require supplementation in view of the so-called electromagnetic retardation. Since according to London's theory of dispersion forces intermolecular interactions are due to electromagnetic interaction, which is propagated with a finite velocity, interactions at sufficiently great distances for which the time of passage of the electromagnetic wave is commensurable with the period of electron rotation, should be weaker than that predicted by London's theory. This was initially pointed out by Overbeek, and calculations of the effect were carried out in 1946 by Casimir and Polder[5]. According to Overbeek[4], for an interaction between two identical molecules the result can be represented by

$$-\frac{\beta}{r^6}\mathrm{f}(p) \quad \text{where} \quad p = \frac{2\pi r}{\lambda}$$

($\lambda = c/\nu$, where c is the velocity of light in vacuum, and ν the frequency in London's law, $\beta = \frac{3}{2}\pi\alpha^2\hbar\nu$, where α is the polarisability of the molecule) where for

$$0 < p < 3, \quad \mathrm{f}(p) = 1.01 - 0.14\,p$$

and for

$$3 < p < \infty, \quad \mathrm{f}(p) = \frac{2.45}{p} - \frac{2.04}{p^2}. \tag{6.12}$$

Calculation of $\Delta\mu$ of a free layer with the aid of (6.12), as the energy of the interaction of a molecule separated by a distance h from a semi-infinite phase of the same substance (with the opposite sign) can be readily carried out by steps analogous to those used in the derivation of (6.10) (Fig. 44). According to (6.12), the integration for $h < \frac{3}{2}\lambda/\pi$ should be performed in two stages: first from h to $r = \frac{3}{2}\lambda/\pi$, corresponding to $p = 3$, according to the law of the interaction of a spherical segment with a molecule:

* Minimum of interaction corresponding to stable equilibrium.

$$-\frac{2\pi\beta}{\bar{v}}\left(1.01 - 0.14\frac{2\pi r}{\lambda}\right)\frac{r-h}{r^5}\,\mathrm{d}r\,,$$

and subsequently for $r = \frac{3}{2}\lambda/\pi$ to $r = \infty$, according to the law:

$$-\frac{2\pi\beta}{\bar{v}}\left(\frac{2.45}{2\pi r}\lambda - \frac{2.04}{4\pi^2 r^2}\lambda^2\right)\frac{r-h}{r^5}\,\mathrm{d}r\,.$$

For $h > \frac{3}{2}\lambda/\pi$,

$$-\frac{2\pi\beta}{\bar{v}}\left(\frac{2.45}{2\pi r}\lambda - \frac{2.04}{4\pi^2 r^2}\lambda^2\right)\frac{r-h}{r^5}\,\mathrm{d}r$$

should be integrated between $r = h$ and $r = \infty$.

Assuming that $p = 2\pi h/\lambda$, the result has the following form: for $0 < p < 3$,

$$\Delta\mu = \frac{A\bar{v}}{6\pi h^3}(1.01 - 0.28p + 14.2\cdot 10^{-3}p^3 - 1.96\cdot 10^{-3}p^4);$$

for $3 < p < \infty$,

$$\Delta\mu = \frac{A\bar{v}}{6\pi h^3}\left(\frac{1.47}{p} - \frac{0.82}{p^2}\right). \qquad (6.13)$$

Such more complete calculations can be also carried out for a layer on a base [eq. (6.11)] using the law of interaction and allowing for the retardation in the case of molecules 1 and 2.

When the law of intermolecular interaction is corrected for the electromagnetic retardation, it yields for large h, ($h \to \infty$), a value of $\Delta\mu$ which is inversely proportional to the fourth power of the distance. However, the extreme case which corresponds to such a simplified law lacks any special meaning, since $\Delta\mu \to 0$ with increasing h, and the corresponding effect on μ is very slight. The case of relatively small h is of a greater interest, since the simple equations (6.10) and (6.11) may then be used for $\Delta\mu$ if λ is sufficiently large.

If two dipolar molecules mutually interact, then in addition to the dispersion interactions considered above, simple dipole [Keesom, (1915–1921)[6]] and induced dipole [Debye, (1920–1921)[7]] interactions will also occur, which vary inversely with the sixth power of the distance. The latter two appear however to play only a minor part in the total interaction between condensed phases which determines $\Delta\mu$, since the dipole interactions are not additive and the overall effect is strongly reduced. For this reason these interactions have so far been neglected[4], even in the special

References p. 207

cases of highly polar molecules (H_2O, NH_3) when the Keesom and Debye components are greater than the London component.

The part played by the entropy component of $\Delta\mu$ of liquid layers, *i.e.* $T(\partial S/\partial n)_{O,T} - T(\partial S_{\infty}/\partial n)_{O,T}$ is still unclear. The only semi-quantitative analysis of this problem (similar to Debye's theory of solids), based on the fact that frequencies of the wave-like movements of thermal origin are limited in a thin layer by its thickness, can be found in some detail in the book by Frenkel'[8]. It was pointed out in Chapter IV that the amplitudes of surface waves for $\sigma > 10$ are very small at normal temperatures, being of the order of molecular dimensions. Contribution of the entropy component to the total energy is thus apparently small and is probably negligible up to monomolecular thicknesses. Frenkel' notes quite rightly that in the case of a rarefied monolayer, the entropy component becomes predominant to such a degree that the usually employed isotherms of "gaseous" monolayers are treated on the basis of this component. This is due to the qualitative change in the nature of thermal molecular motion which takes place on transition from the coordinated movement of molecules without any change in their mutual distances in dense mono- or multimolecular layers to the free translational movement on the surface of the rarefied monolayer (non-stretched films). Solution of these problems, which is highly important for the development of a consistent general theory of adsorption, that would be valid for both mono- and multimolecular adsorption, is still however to be found*.

Let us now consider one more special but highly important case of thin layers of dilute aqueous electrolytes. If a considerable potential jump (φ_0 potential) exists at the phase surface of such a layer, then a diffuse electric layer will be present in the film. As long as the film is sufficiently thick, these layers in the film do not differ in the least from analogous layers in a film of infinite thickness, and they do not influence μ. At smaller values of h the diffuse layers are compressed, and the energy of their deformation results in an alteration of the free energy of the liquid film. Let us analyse the simple case of a symmetrical film with identical φ_0 potentials and correspondingly containing two identical diffuse layers. Two such films are shown in Fig. 45: (a) is a film of great thickness, in which the diffuse layers have been fully developed, so that the internal

* Evaluation of the entropy component of the interaction between two apparently solid phases may be found in the bibliography[9] carried out from the standpoint of the macroscopic theory (see further).

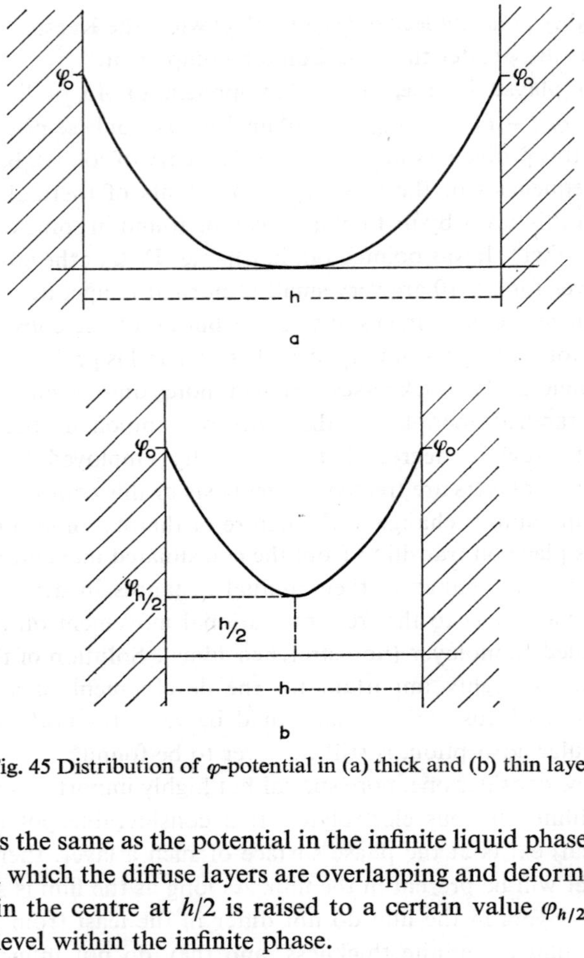

Fig. 45 Distribution of φ-potential in (a) thick and (b) thin layers.

potential is the same as the potential in the infinite liquid phase, whilst (b) is a film in which the diffuse layers are overlapping and deformed, and the potential in the centre at $h/2$ is raised to a certain value $\varphi_{h/2}$ above the potential level within the infinite phase.

Let us calculate the corresponding change of the chemical potential, $\Delta\mu_{el}$, caused by the compression of the diffuse electric layers, making use of the following model.

According to the principles of electrostatics, an attractive force

$$f = \frac{2\pi}{\varepsilon} e^2 \tag{6.14}$$

acts on every cm^2 between the plates of a plane condenser, assuming that the plates are infinite and carry a charge of e per cm^2. It follows that the outer plate of the electrical double layer presses on the bulk phase with this force, increasing in it the pressure due to van der Waals' attractive forces. According to (6.2), this results in an increase of chemical potential of the phase. If the surface charge e is altered, a change will also take place in the chemical potential, as can be clearly seen in electrocapillary phenomena. According to (6.14), this effect is independent of the sign of the external charge. The chemical potential will therefore increase if the charging is carried out about the point of zero charge, irrespective of whether the phase is question is charged positively or negatively.

It should be emphasised that this electrostatic component of $\Delta\mu$ is only a part of the complete electrochemical component of $\Delta\mu$. If for example a change in the charge is associated with a change in the composition, as is generally the case, the total change in chemical potential is due not only to the electrostatic but also to another, "osmotic", component.

When the film is thinner and the diffuse layers are correspondingly compressed, the double layer charges undergo a decrease. According to (6.14) these changes will result in a corresponding change of pressure, and consequently of chemical potential. In the particularly simple case of a symmetric film, with sufficiently small degrees of overlap in the electrical diffuse layers, when their overlapping and deformation occurs predominantly in the "tails" without affecting the structure near the phase surface, it can be assumed that the φ_0 potential is constant.

According to (6.14) the pressure change in the film, caused solely by such deformation of the diffuse electrical layers, *i.e.* the electrostatic component of the disjoining pressure is:

$$\Pi_{\text{el}} = \frac{2\pi}{\varepsilon}(e_\infty^2 - e_h^2) \tag{6.15}$$

where e_∞ is the charge of the undeformed diffuse layer, and e_h is the charge of the layer in a film of thickness h. These surface charges are respectively

$$e_\infty = \int_{x=0}^{x=\infty} \rho\,\mathrm{d}x \quad \text{and} \quad e_h = \int_{x=0}^{x=h/2} \rho\,\mathrm{d}x^{*}.$$

* Each cm² of a film with an area of 2 cm² corresponds to a charge $2\int_0^\infty \rho\,\mathrm{d}x$ or $2\int_0^{h/2} \rho\,\mathrm{d}x$ (for thick and thin layers respectively), so that the charge per unit of area is $\int_0^\infty \rho\,\mathrm{d}x$ or $\int_0^{h/2} \rho\,\mathrm{d}x$.

Making use of the Poisson equation $d^2\varphi/dx^2 = -4\pi\rho/\varepsilon$, we obtain after integration

$$e_\infty = \frac{\varepsilon}{4\pi}\left(\frac{d\varphi}{dx}\right)_{0,\infty} \quad \text{and} \quad e_h = \frac{\varepsilon}{4\pi}\left(\frac{d\varphi}{dx}\right)_{0,h}$$

where the derivatives $d\varphi/dx$ relate respectively to the initial part of the diffuse layer (subscript "0"), to the undeformed (subscript "∞") and to the compressed (subscript "h") layers. The derivatives which correspond to the upper limits of integration: $(d\varphi/dx)_{\infty,\infty}$ and $(d\varphi/dx)_{h/2,h}$ are equal to zero, the latter because at $h/2$ the function $\varphi(x)$ shows an obvious minimum (Fig. 45).

By using the expression for a binary electrolyte (5.28) from the Gouy–Chapman's theory, we obtain

$$\frac{d^2\varphi}{dx^2} = -\frac{4\pi\rho}{\varepsilon} = \frac{8\pi Fz}{\varepsilon} c \sinh\frac{Fz\varphi}{RT}.$$

After integration from $x = 0$ to $x = \infty$ ($\varphi = \varphi_0$ and $\varphi = 0$) and from $x = 0$ to $x = h/2$ ($\varphi = \varphi_0$ and $\varphi = \varphi_{h/2}$) respectively, we obtain

$$\left(\frac{d\varphi}{dx}\right)^2_{0,\infty} = \frac{16\pi cRT}{\varepsilon}\left(\cosh\frac{Fz\varphi_0}{RT} - 1\right)$$

and

$$\left(\frac{d\varphi}{dx}\right)^2_{0,h} = \frac{16\pi cRT}{\varepsilon}\left(\cosh\frac{Fz\varphi_0}{RT} - \cosh\frac{Fz\varphi_{h/2}}{RT}\right)$$

which, after substitution into (6.15), yields

$$\Pi_{el} = 2cRT\left(\cosh\frac{Fz\varphi_{h/2}}{RT} - 1\right). \tag{6.16}$$

The final result is obtained after elimination of $\varphi_{h/2}$ which is obviously a function of φ_0. This function can be obtained with relative ease if we limit ourselves to an approximation for very low values of $\varphi_{h/2}$, *i.e.* to the case of small deformations and overlaps of the diffuse electric layers. Under these conditions $\varphi_{h/2}$ is approximately twice the φ-potential at the distance $h/2$, which would arise due to each of the two diffuse layers in the absence of the other layer. For this approximation we have, according to (5.36)

References p. 207

$$\varphi_{h/2} = \frac{2RT}{zF} 4\gamma \, e^{-\varkappa h/2}.$$

In (6.16)

$$\cosh(Fz\varphi_{h/2}/RT) = \frac{1}{2}\left[\exp\left(\frac{Fz\varphi_{h/2}}{RT}\right) + \exp\left(-\frac{Fz\varphi_{h/2}}{RT}\right)\right]$$

is expanded into a power series of $Fz\varphi_{h/2}/RT$ according to the formula $e^x \approx (1+x+x^2/2)$ (the second terms in cosh cancel out), to give

$$\Pi_{el} = cRT\left(\frac{Fz\varphi_{h/2}}{RT}\right)^2.$$

After substituting for $\varphi_{h/2}$ this becomes

$$\Pi_{el} = 64cRT\gamma^2 \, e^{-\varkappa h} \tag{6.17}$$

or, corresponding to (6.2) for the electrostatic component of the excess chemical potential of a thin layer of electrolyte solution,

$$\Delta\mu_{el} = -64cRT\gamma^2 \bar{v}_1 \, e^{-\varkappa h}. \tag{6.18}$$

In spite of the simplicity and approximate nature of eqns. (6.17) and (6.18), they are nevertheless sufficient for an explanation of the phenomenon. The theory of the effect in question evolved in connection with attempts to explain the stability of lyophobic colloids and was introduced by Deryagin in 1937[10] and by Bergmann, Löw–Beer and Zocher in 1938[11]. A more complete form was developed by Deryagin and Landau (1941)[12]. Most of the further refinements of the theory, which may be found in the bibliography[4,13] consist of more accurate presentations which do not alter the nature of the phenomenon as discussed above*. Among the later supplements the most interesting is probably the case of an asymmetric layer with various values of the φ_0-potential for both surfaces [Deryagin, Levich (1954)[14]]. The physically new elements, which have not yet been incorporated into the theory are the allowance for Stern's correction to the theory of the double layer, and allowance for the effect of van der Waals' forces on the distribution of charges in the deformed diffuse layers in a thin film

* Π_{el} is generally calculated by other methods without using the model employed by ourselves. Since this effect is of purely thermodynamic origin, its mechanism is immaterial and various methods of calculation are possible. The present author believes, however, that the use of a model as above entails, within the framework of thermodynamic interpretation, a certain advantage of greater clarity.

of an electrolyte solution. This points definitely to a mutual dependence of the van der Waals' and the electrostatic components of $\Delta\mu$.

The above argument brings us to the most complex case of a "non-one-component" film, *i.e.* a film in which the components are present in commensurable quantities, and where van der Waals' forces can affect considerable concentration changes in the film as compared with the concentration in a layer of infinite thickness. The peculiarity of this highly important case which has recently been studied by Deryagin and Martynov (1962)[15], lies in the following. According to (6.10), $\Delta\mu$ of a thin structureless free one-component layer is invariably positive, corresponding to a negative disjoining pressure and to thermodynamic instability, provided that the film does not contain diffuse electric layers which always give rise to a negative component $-\Delta\mu_{el}$. In the case of a "multicomponent" film, one may also expect positive values of Π, *i.e.* the formation of stable films as a result of concentration changes and corresponding osmotic effects. This has been so far considered by many authors to be impossible in principle.

Relatively recently Lifshits (1954)[16] put forward a new, more general theory of the van der Waals' component of the attraction between two semi-infinite phases separated by a plane-parallel gap; this theory was later extended by Dzyaloshinskii, Lifshits and Pitaevskii (1959)[17] to the general case of a thin layer compressed between two different semi-infinite phases. In this theory $\Delta\mu$ is regarded as a result of the interaction between fluctuating electromagnetic fields which extend beyon dthe phases, and is analysed on the basis of a method developed by Rytov (1953).[18] This theory is extremely complex and, in its latest form, makes use of quantum electrodynamics; it therefore cannot be dealt with in the present book. Its final equations contain optical frequency functions for the various phases, which have not as yet been investigated. For the simplest boundary case of a sufficiently thin free film, it yields a $\Delta\mu$ which varies inversely as the third power of thickness; for the interaction of two molecules the theory predicts that the energy of interaction varies inversely with the sixth power of distance. These agreements with the molecular treatment of the problem referred to before, appear to indicate that consideration of the fluctuating electromagnetic field of the phase (considered from a very general point of view without making an allowance for its molecular structure, as a kind of generation of such a fluctuating field) used in the theory of Lifshits, provides a more general expression of London's model representation of

References p. 207

fluctuating dipoles (and correspondingly of the fluctuating fields) of neutral atoms. It is interesting to note that, for the case of a thick layer, Lifshits' theory predicts an inverse proportionality of $\Delta\mu$ to the fourth power of distance, and thus apparently also includes a correction for the time lag in a more general form. (For the interaction between two molecules separated by a large distance it is correspondingly obtained that the energy varies inversely as the seventh power of the distance.)

Although Lifshits' theory has so far been confined to one-component structureless layers and deals therefore with the same systems as the molecular theory, its use is undoubtedly of great interest. In view of its more general character its principles are expounded in a more accessible manner in Deryagin's paper[9], and in greater detail in Abrikosov's book[24].

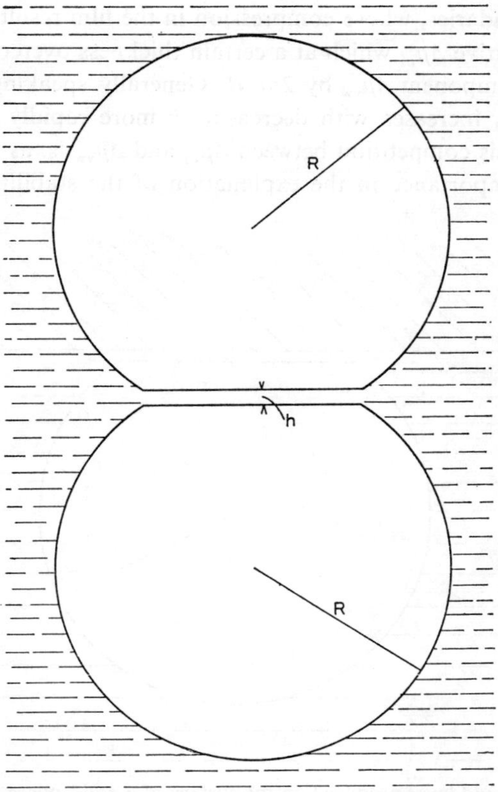

Fig. 46. Free liquid layer formed between two identical bubbles.

Let us now analyse briefly from these vantage points the problem of the stability of thin liquid layers, making use for the sake of clarity of the example of a free layer formed between two identical bubbles brought together in a liquid (Fig. 46). According to the Thomson–Gibbs equation, the chemical potential in the bubbles is reduced by $2\sigma\bar{v}/R$ and correspondingly the vapour pressure is $p_r < p_\infty$. The thin plane-parallel layer h might now be in equilibrium, had its chemical potential been reduced by the same magnitude. It has been shown, however, that the μ_h of such a layer increases above μ_∞ of the large phase when the thickness is decreased, owing to van der Waals' forces, and consequently such free films cannot be in equilibrium with the bubbles. Equilibrium would only be possible in the case of highly polar liquids (H_2O, NH_3) at low electrolyte concentrations, because under these conditions diffuse electric layers arise at phase boundaries, whose compression in the film results in the formation of a negative $\Delta\mu_{el}$ which at a certain thickness overcompensates van der Waals' component $\Delta\mu_{vw}$ by $2\sigma\bar{v}/R$. Generally speaking this is possible since $\Delta\mu_{el}$ increases with decreasing h more rapidly (exponentially) than $\Delta\mu_{vw}$. This competition between $\Delta\mu_{el}$ and $\Delta\mu_{vw}$ is, as will be seen, of the utmost importance in the explanation of the stability of lyophobic

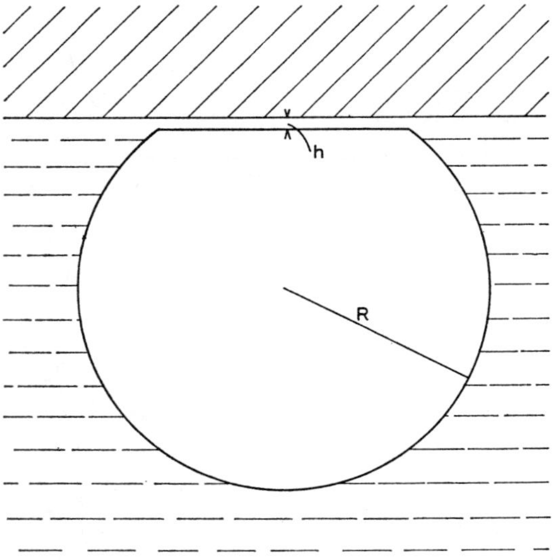

Fig. 47. Thin liquid layer produced on the surface of a solid phase when a bubble is pressed against it.

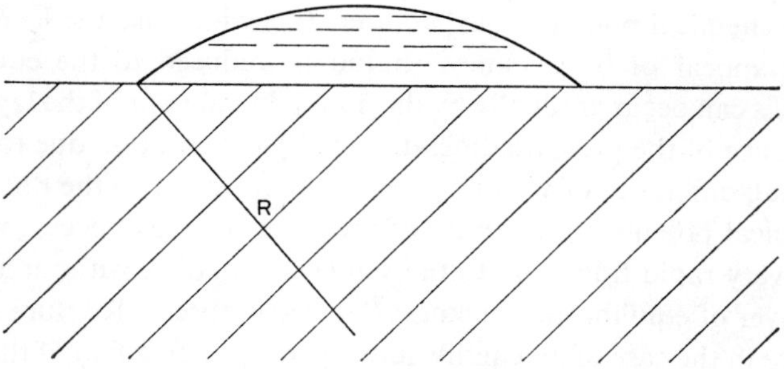

Fig. 48. A drop on the surface of a solid.

colloids. However, for most liquids $\Delta\mu_{el} = 0$, and the free layer cannot coexist in equilibrium with a large mass of the liquid.

Fig. 47 shows the example of a thin liquid layer produced on the surface of a solid phase when a bubble is pressed against it. In this case equilibrium can be achieved not only due to $\Delta\mu_{el}$, but also due to $\Delta\mu_{vw}$ since the latter can be negative [eqn. (6.11)].

The chemical potential of a drop on a solid or liquid surface (Fig. 48) is increased by $2\sigma\bar{v}/R$. The drop could therefore exist in equilibrium with a thin layer on the surface of a solid whose chemical potential is similarly increased, and this is possible for $\Delta\mu_{vw}$ in the case of a layer supported on a base. If however $\Delta\mu$ of the layer varies inversely as the third power of the thickness (6.11) or, in general, increases monotonically with decreasing h, this equilibrium would be unstable. As the thickness of the layer is decreased (passage of liquid from the layer into the drop) its $\Delta\mu$ is increased, while $\Delta\mu$ in the drop is reduced, leading to further passage of the liquid from the layer into the drop and to a progressive increase of the difference between their chemical potentials, until the layer disappears. Equilibrium is impossible when the $\Delta\mu$ of the layer is negative, and the drop should spread out into a layer. In other words negative values of $\Delta\mu$ in a layer supported on a base correspond to complete wetting of the base by the liquid, whilst incomplete wetting indicates that $\Delta\mu$ of the layer is positive, at least for layers which are thicker than the equilibrium layer. Equilibrium between the layer and the drop is only possible if the sign of $\Delta\mu$ is changed, which is not foreseen by the elementary theory for $\Delta\mu_{vw}$, but which can take place for the combination of $\Delta\mu_{vw}$ with $\Delta\mu_{el}$.

It is thus clear that there are possible positions at which the layer is not in equilibrium with the bulk phase when its chemical potential is higher than the chemical potential of the latter. In such a case the layer should either disappear or its thickness should be reduced to the equilibrium value. This can occur gradually by the liquid flowing out of the layer under the influence of the pressure difference, or by evaporation due to the corresponding difference of vapour pressures (equivalent to the difference in the chemical potentials). Rupture of the layer may also occur, with subsequent very rapid transition to the stable state, either without a layer or with a layer of equilibrium thickness ("semi-rupture"). Rupture may also take place in the case of an equilibrium layer, *e.g.* a free film, if the state of equilibrium is relatively less stable than that arising after rupture.

These kinetic problems will be described in connection with the experimental investigations of thin layers.

2. Experimental Investigations of Thin Layers

These investigations were to a large extent initiated and developed by Deryagin *et al.*

The first, crucial problem consists of demonstrating that in spite of their very short range the intermolecular attraction forces give rise to measurable effects (*e.g.* Π) at inter-phase distances which considerably exceed molecular dimensions. The order of magnitude of Π can be estimated theoretically[3] in the following manner. If the phase is divided into two semi-infinite parts which are then moved reversibly to a large distance from each other, then the work done in overcoming the attraction forces between them will be in agreement with (6.2) and (6.10).

$$-\int_{r_0}^{\infty} \Pi \, \mathrm{d}h = \int_{r_0}^{\infty} \frac{K}{h^3} \mathrm{d}h = \frac{K}{2r_0^2} \text{ per cm}^2 ,$$

where r_0 is the minimum distance, of the order of a molecular diameter, at which the rupture starts. The free energy (per 1 cm^2 of the gap) of the system will then increase by 2σ (where σ is the free surface energy). Consequently

$$K = 4\sigma r_0^2 . \tag{6.19}$$

For water ($\sigma \approx 10^2$, $r_0 \approx 2 \cdot 10^{-8}$) this yields for $\Pi = -K/h^3$ at $h = 1000$ Å, the value $\Pi \approx |10^2|$, *i.e.* an easily measurable magnitude. It was however mentioned in the derivation of (6.11) that the use of this method of calculation for small values of h is of doubtful validity. Therefore $\int_{r_0}^{\infty} \Pi \, \mathrm{d}h$,

References p. 207

whose main part is formed at small values of h, close to r_0, cannot be considered as reliable when eqn. (6.11) is used and (6.19) cannot unfortunately be used with full confidence as a proof of the effectiveness of the long-distance effect.

In this connection it was very important to demonstrate the existence of noticeable interaction forces between condensed phases. This was first carried out successfully by Deryagin and Abrikosova (1953–1954)[19] by means of a sensitive beam balance fitted with an electromagnetic appliance operating on the feedback principle, and thus combining the necessary stability (aperiodicity) with a high sensitivity. The basic measurements were carried out for the attraction between a convex quartz lens and a plane quartz plate, in air and in vacuum. This special form of the gap was chosen in connection with the unevenness of solid surfaces and the consequent impossibility of creating a regular plane-parallel gap between them. The minimum distance for each gap was measured very accurately by the interferometric method, in reflected monochromatic light. For this form of

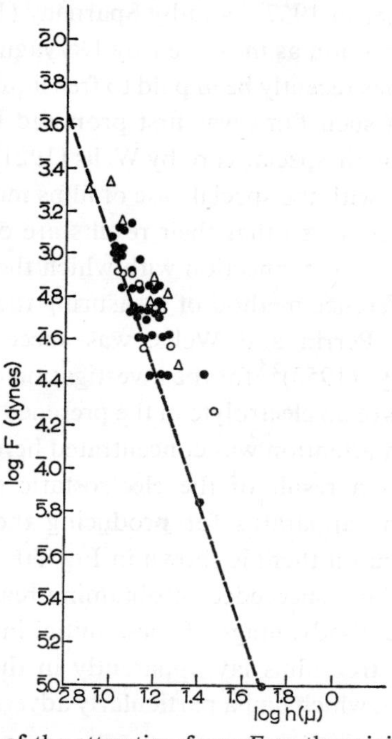

Fig. 49. Dependence of the attraction force F on the minimum distance h between a convex quartz lens and a plane quartz plate.

gap, the forces of interaction should decrease by one order of magnitude more steeply than for a plane-parallel gap, and the interaction force should become even weaker. All this resulted in an unavoidable and considerable scatter of results. The experimental data of Deryagin and Abrikosova are shown in log–log scale on Fig. 49. The theoretical dependence for this case (hemisphere/plane), as calculated approximately on the basis of Lifshits' theory* is shown by the dotted line. This relationship is close to that obtained with the aid of the theory of intermolecular attraction, in which electromagnetic retardation is allowed for. For details of this complex experiment and of its interpretation, see ref. 9. A special prerequisite in this case is the elimination of surface charge which easily accumulates on the surface of the insulator (quartz). Analogous measurements carried out almost simultaneously (1952) by Overbeek and Sparnaay yielded (probably due to an incomplete elimination of the surface charge) values of attraction forces which were 3 to 4 orders of magnitude greater than those predicted by the theory. The experimental check of these measurements, carried out by Kitchener and Prosser in 1957[20] and by Sparnaay (1960)[21], confirmed the correct order of attraction as measured by Deryagin and Abrikosova.

Particular attention has recently been paid to free liquid films. Investigation of the thickness of such films was first proposed by Perrin, and the work was carried out with special care by Wells (1921).[22] Unfortunately these authors dealt only with the special case of films made from relatively concentrated soap solutions, so that their results are only of interest for the theory of soap foams, in connection with which they will be analysed later. The optical interference method of measuring the thickness of such films, put forward by Perrin and Wells, was successfully applied by Deryagin and Titievskaya (1953)[23] for the investigation of films made from dilute aqueous solutions of an electrolyte in the presence of various surface-active stabilisers. Main attention was concentrated here on the formation of equilibrium films as a result of the electrostatic component of the disjoining pressure. The apparatus for producing the films at a given external pressure exerted on them is shown in Fig. 50. Making use of the theory for Π_{el}, the authors succeeded in obtaining reasonable values for the surface charge. The disadvantage of these initial investigations of the disjoining pressures of free films lay apparently in the low accuracy of thickness determination, which had a particularly adverse effect in the case

* As demonstrated in ref.[9], the optical functions included in the Lifshits' theory can be evaluated for the special case of quartz.

References p. 207

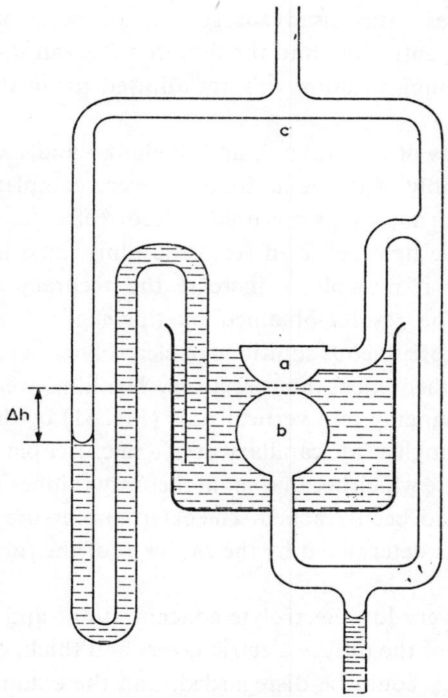

Fig. 50. Deryagin and Titievskaya's apparatus for obtaining free liquid films at a given external pressure ($\Delta h\rho g$).

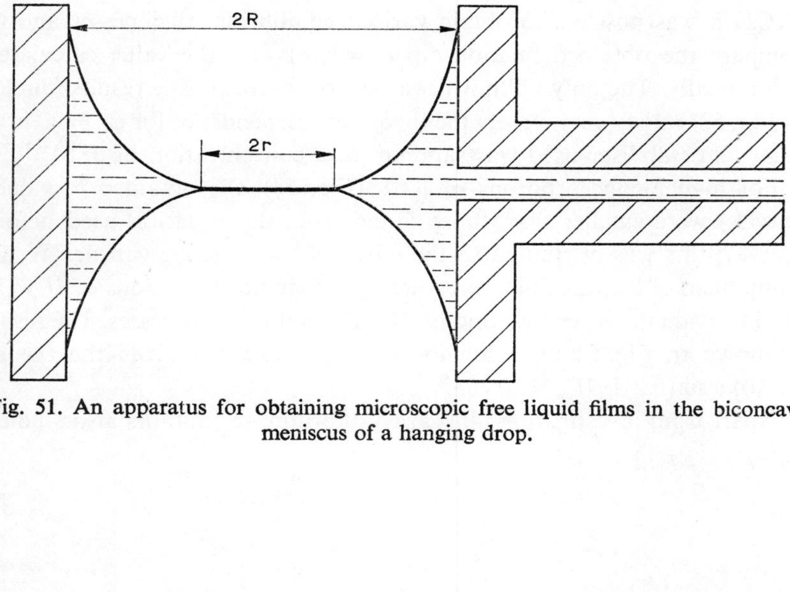

Fig. 51. An apparatus for obtaining microscopic free liquid films in the biconcave meniscus of a hanging drop.

of small thicknesses; other disadvantages were the occasional uncertainties in electrolyte concentration, and the fact that the van der Waals component of the disjoining pressure was not allowed for in the interpretation of results.

The experiments of Sheludko[25], and Sheludko and Ekserova[26], which were started roughly at the same time but were completed later, and in which a registering apparatus was used to record changes in the brightness of monochromatic light reflected from the film whilst its thickness was decreasing, made it possible to increase the accuracy of the thickness measurements. The results obtained by this method pertained to the equilibrium films of aqueous solutions of electrolytes, and lent themselves more reliably to theoretical interpretation. The films were produced in a biconcave drop hanging in a vertical tube (Fig. 51) by drawing away the liquid from it through a side capillary until a microscopic circular film was formed in the centre which became spontaneously thinner until the equilibrium thickness had been reached. The external pressure under which the film remained was determined by the radius R of the tube as approximately $2\sigma/R = P_0$.

In the case of very low electrolyte concentrations and correspondingly large thicknesses of the diffuse electric layers and thick, equilibrium films (thickness h_r), Π_{vw} could be disregarded, and the equilibrium condition could be written as

$$\Pi_{\text{el}} = P_0 = 64cRT\gamma^2 \text{e}^{-\varkappa h}$$

in accordance with (6.17). By varying the concentration of the electrolyte (KCl) it was possible to obtain various equilibrium thicknesses and to compare the obtained relationship $h_r = h_r(c)$ with the value calculated theoretically. The only unknown was φ_0 in $\gamma = \gamma(\varphi_0)$. The results (circles in Fig. 52) satisfy very closely the theoretical dependence for $\varphi_0 = 90$ mV. (The film stabiliser used was saponin, in a concentration of $5 \cdot 10^{-4}$%.)

For higher concentrations of KCl, (Fig. 53), the obtained h_r values (circles) were smaller than those found from the equation used before (curve); this was attributed to the effect of the negative van der Waals component of the disjoining pressure. The obtained deviations of Π were used to evaluate the corresponding Π_{vw} for various thicknesses. The result is shown in Fig. 54 on a log–log scale. As expected from the theory [(6.10) and (6.2)], $\Pi_{\text{vw}} = -K/h^3$.

Apart from investigating films at equilibrium, the authors also studied

References p. 207

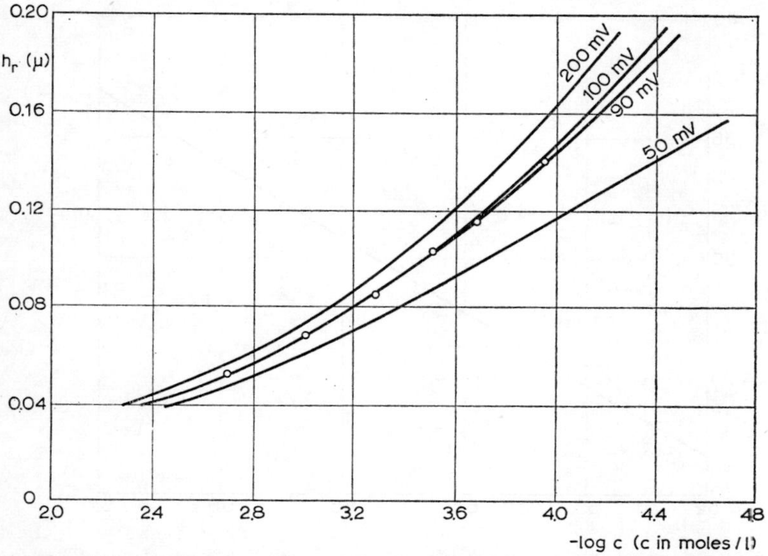

Fig. 52. Dependence of the equilibrium thickness h_r on small concentrations c of electrolyte (KCl) for free liquid films, calculated for four values of φ_0 (—). Experimental data are shown by circles.

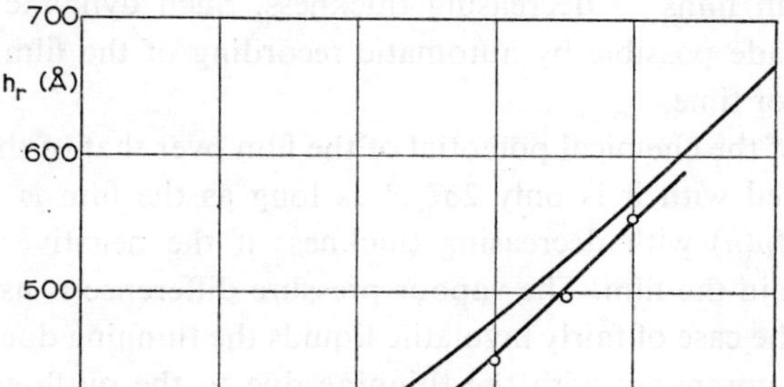

Fig. 53. Dependence of the equilibrium thickness h_r on high electrolyte (KCl) concentrations c, for free liquid films. Experimental data are shown by circles, and the upper curve was calculated from formulae for Π_{el}.

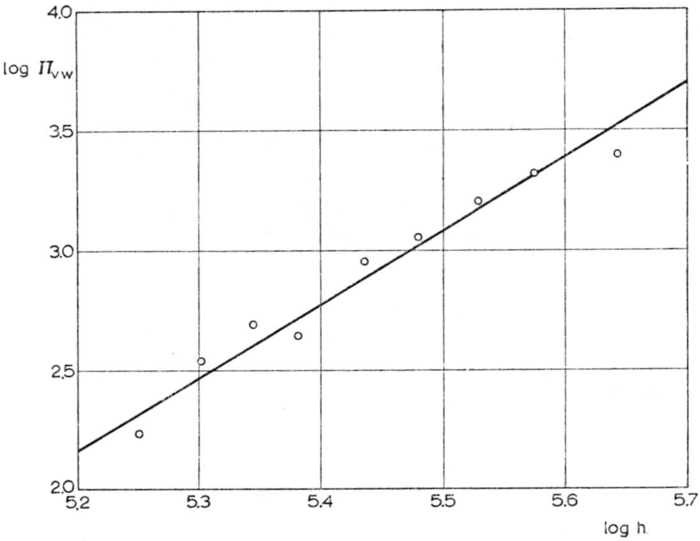

Fig. 54. Isotherm of the van der Waals' component of the disjoining pressure $\Pi_{vw}(h)$ for free liquid films of aqueous solutions of KCl.

non-equilibrium films of decreasing thickness. Such dynamic measurements were made possible by automatic recording of the film thickness over a period of time.

The excess of the chemical potential of the film over that of the concave liquid connected with it is only $2\sigma\bar{v}/R$ as long as the film is thick, but increases by $\Delta\mu(h)$ with decreasing thickness if the negative disjoining pressure arises in the film. The vapour pressure difference is usually very slight, and in the case of fairly involatile liquids the thinning due to evaporation is slow compared with the thinning due to the outflow of liquid from the film into the meniscus as a result of pressure difference $P = 2\sigma/R - \Pi$. If the hydrodynamic law of film thinning is known it becomes possible, by studying the rate of thinning, to calculate the corresponding pressure difference and Π as a function of h, *i.e.* to obtain the isotherm $\Pi = \Pi(h)$ or $\Delta\mu = \Delta\mu(h)$.

Experiments aimed at the determination of the hydrodynamic law of thinning, at first with relatively thick films, by measuring the film thickness as a function of the thinning time from its electrical resistance [Sheludko, Dessimirov, Nikolov (1954)[27]] showed that a direct proportionality exists between $1/h^2$ and the thinning time t. Further measurements, this time with microscopic films [Sheludko (1957)[28]] confirmed this relationship in the form of Reynolds' law

$$\frac{\mathrm{d}(1/h^2)}{\mathrm{d}t} = aP, \qquad (6.20)$$

where P is the required pressure difference ($P = P_0 - \Pi$), and $a = 4/3\eta r^2$ (η is the viscosity of the liquid in the film, and r is its radius). This law is obeyed for sufficiently thin ($h \leqq 1000$ Å) plane-parallel films*. This result is rather unexpected since Reynolds' law was derived for the thinning of a circular liquid film compressed between two solid surfaces, when the flow rate in the surface was equal to zero. The absence of flow in the surface of free films could be explained on the basis of ideas used by Frumkin and analysed in detail by Levich[29]. The surface-active component which must be added to the solution to obtain the film at all (as otherwise an instantaneous rupture occurs when the surfaces of the biconcave drop touch one another) is mainly concentrated in the surface. As the film becomes thinner, the surface-active agent is carried off by the liquid flowing symmetrically from the centre towards the periphery of the film, so that the adsorbed layer is "stretched" and rarefied, and a surface concentration gradient arises in it. This results in the production of a corresponding two-dimensional pressure directed against the flow. Given a sufficient degree of stretching (assuming that the amount of the surface-active component is sufficient) the forces which cause the drainage of the adsorption layer and the reverse surface pressure become fully balanced, and the stretched surface layer attains a stationary value. More detailed investigations of this phenomenon[25] have borne out the correctness of this explanation.

Making use of the relationship (6.20) the present author obtained (1958)[31] the isotherm $\Pi = \Pi(h)$ for aniline (Fig. 55) and for aqueous films with a medium content of electrolyte, when commensurable Π_{vw} and Π_{el} are active in the film, for the range prior to attaining the equilibrium thickness (Fig. 56). Finally, measurements were also carried out with aqueous

* The thicker films are convex in the centre. A hydrodynamic theory of the thinning of films, in which the form of the film profile would be determined by the conditions of flow, is under development by Mysels and Frankel.[30] The preliminary communication of these authors shows that they have succeeded theoretically in obtaining a convex profile. The extension of the theory to films of large (non-microscopic) dimensions will probably prove to be a highly complicated task since to judge from the experimental data, the conditions of flow in large films are extremely complex.

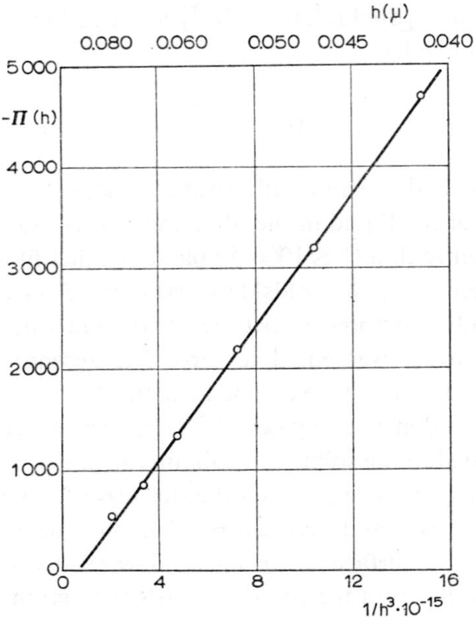

Fig. 55. Isotherm of the disjoining pressure $\Pi(h)$ for free aniline films, obtained from the rate of their thinning.

films containing a considerable excess of electrolyte (which suppresses the diffuse electric layers) for a wide range of thicknesses, where Π_{vw} manifests itself in the "pure form"[25] (Fig. 57).

Recently Duyvis (1962)[32] repeated those parts of the above investigations which pertained to the equilibrium films of aqueous electrolyte solutions, using measuring cells which were a successful combination of the previous type with a modified method of interferometric thickness measurement. Duyvis' results, obtained with a large number of surface-active components as stabilisers, are in agreement with those of Sheludko and Ekserova, As a result of all these studies, it is found that for water at $\Pi = -K/h^3$, K_{H_2O} is $\approx 4 \cdot 10^{-14}$, or that correspondingly $A_{H_2O} \approx 7.5 \cdot 10^{-13}$, whilst $K_{aniline} = 3.5 \cdot 10^{-13}$ ($A_{aniline} = 6.6 \cdot 10^{-12}$).

The data relating to the van der Waals component Π for aniline and water are characterised by a direct proportionality between Π and $1/h^3$, which corresponds to the theory when the effect of electromagnetic retardation is discounted. If it is assumed as usual[4] that the length oı a London wave is 1000 Å, then this result is inexplicable: in the range of the measured

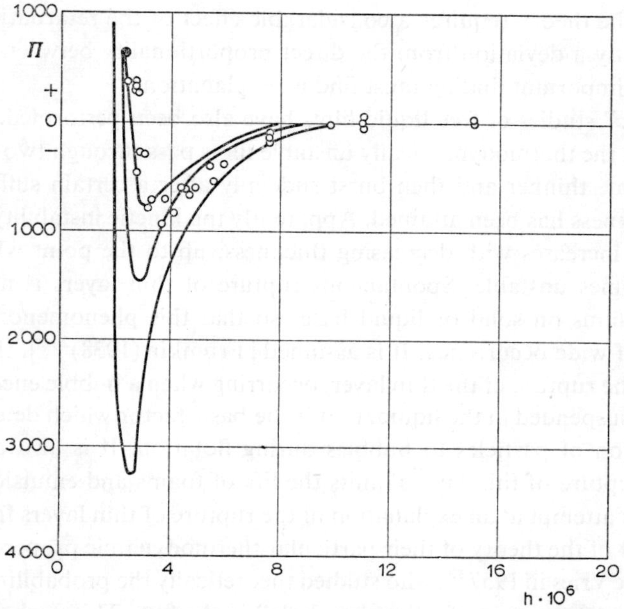

Fig. 56. Isotherm of the disjoining pressure $\Pi(h)$ for free films of electrolyte (KCl) solutions, up to the equilibrium (black dot). The experimental data are shown by circles.

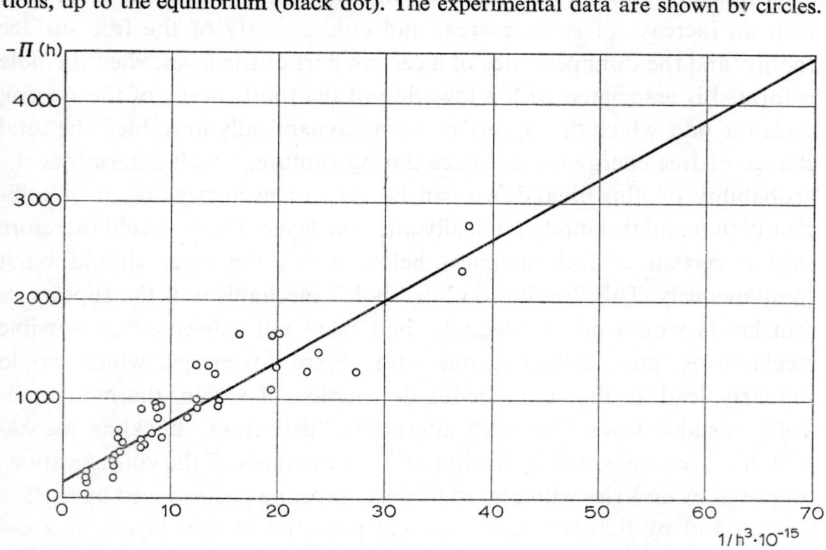

Fig. 57. Isotherm of the disjoining pressure $\Pi(h)$ for a relatively concentrated (0.1 M) solution of KCl.

thickness the theory requires a considerable effect of the retardation and consequently a deviation from the direct proportionality between Π and $1/h^3$. This important finding must find an explanation.

"Kinetic" studies of free liquid films have also been performed. It was found that the thermodynamically unstable films pass through two stages: they become thinner and then burst suddenly after a certain sufficiently small thickness has been attained. Apparently the kinetic instability of the thin layer increases with decreasing thickness, up to the point when the layer becomes unstable. Spontaneous rupture of thin layers is also observed in films on solid or liquid bases so that this phenomenon is apparently of wide occurrence. It is assumed [Frumkin (1938)[2,4]], that it is precisely the rupture of the thin layer, occurring when a bubble encounters a particle suspended in the liquid, that is the basic factor which determines the adhesion of particles to bubbles during flotation. It is also obvious that the rupture of thin layers limits the life of foams and emulsions.

The first attempt at an explanation of the rupture of thin layers from the standpoint of the theory of their particular thermodynamic properties was made by de Vries in 1957[33], who studied theoretically the probability of the spontaneous formation of a "nucleus hole" in the film. This probability is determined by two opposing factors: the formation of a hole is associated with an increase of surface area, and consequently of the free surface energy, and the disappearance of a certain part of the layer when the hole is formed is associated with a lowering of the total energy of the system, since the part which disappears is thermodynamically unstable. The total change of free energy taking place during rupture, which determines the probability of film breakdown can be zero or even negative in a sufficiently thin and thermodynamically unstable layer. There should therefore exist a certain critical thickness below which the layer should burst spontaneously. This "cavitation" or "hole" mechanism of the rupture of thin layers would prove adequate, had there not existed other possible mechanisms, more advantageous with regard to energy, which would similarly lead to the spontaneous destruction of a thin, thermodynamically unstable layer. One such alternative "deflection" buckling mechanism has been suggested by Sheludko[25], on the basis of the considerations proposed by de Vries with regard to the competing parts played by surface tension and by the increased chemical potential in thin layers. If a deflection of the surface and a corresponding local thinning is produced in a thin layer, then in the deflected section there arises on the one hand a

References p. 207

capillary pressure directed against the deflection and tending to cancel it, and, on the other hand, if the layer is thermodynamically unstable, there will arise at the thinner spot an excess of the negative disjoining pressure directed towards thinning, *i.e.* tending to deepen the produced deflection. It is clear that in thinner, thermodynamically less stable layers, the latter effect will be greater and the instability will increase. As soon as the total thickness of the layer becomes sufficiently small, so that the disjoining (negative) pressure begins to predominate over the capillary pressure, the form of the layer will become unstable and the minimum deflection (local thinning) should develop progressively. This should lead either to the rupture of the layer or to a sudden transition to a lesser, thermodynamically or kinetically stable thickness, if such a thickness is produced and if the layer can avoid rupture during this sudden change of thickness. These qualitative considerations yielded the following relation for the critical rupture thickness[25]:

$$h_{cr} = \sqrt[4]{\frac{3K\lambda^2}{64\sigma}}, \tag{6.21}$$

where K is the constant $\Pi = -K/h^3$, and λ is the wavelength of the film surface deflection. It follows that deflections with various $\lambda < r$ (where r is the film radius) may occur in the film, but not all of them are equally probable. Small values of λ are advantageous because they are associated with the transfer of liquid in the film over small distances, and this transfer is associated with considerable friction (6.20); on the other hand large values of λ are advantageous because, for the same deflection, they generate a lesser curvature, *i.e.* they are less encumbered by capillary forces. In view of the occurrence of these two factors, the existence of a certain most probable value of λ may be assumed, which should correspond to a certain, most frequently observed value of critical thickness. Experimental investigations of the critical thicknesses of rupture (Fig. 58, aniline), carried out by Ekserova in 1961[25], showed in fact that the values of the critical thickness of rupture are scattered about a certain most probable value which, in accordance with (6.21), corresponds to $\lambda \sim 0.1r$; this is rather lower for aniline and higher for aqueous solutions. These and some other results[25], confirm the above described mechanism of the rupture of thin layers. On the basis of this mechanism Sheludko (1963)[34] put forward a version of the theory of the elementary act of flotation.

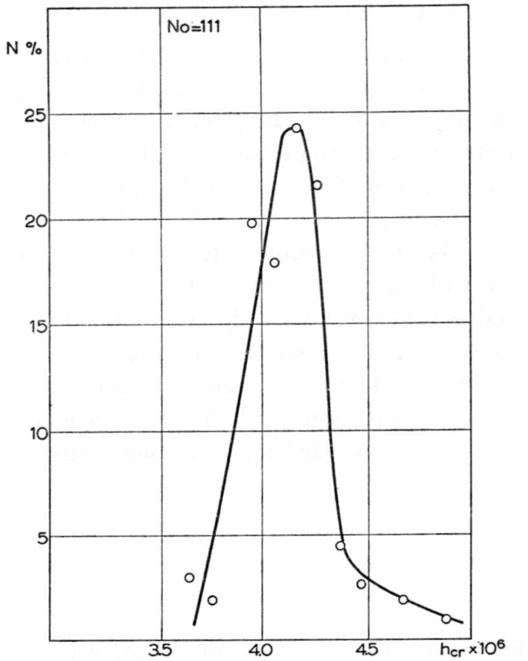

Fig. 58. Distribution of critical rupture thickness h_{cr} in relation to the number of measurements N, the total number of measurements being $N_0 = 111$, for free aniline films of diameter $\approx 2 \cdot 10^{-2}$ cm. The most frequent value of h_{cr} is 420 Å.

The results of such studies of free thin layers suggest very strongly that this new field is promising, and that it will be developed further.

The most complex case which will be described here is that of thin liquid layers supported on a solid or liquid base. The method used in the outstanding work of Deryagin and Kusakov (1937)[35] initiated the systematic quantitative studies of thin liquid layers of controlled thickness. An air bubble held by a special holder is either pressed against the studied surface-base until a plane-parallel equilibrium film is formed on it, or it is placed under a transparent plate to which it is pressed by buoyancy. The thickness of the equilibrium layer, if such a layer is formed, is measured by an interferometric method in reflected monochromatic light. It is interesting to note that in the degree thesis of Kusakov (1941)[36], which unfortunately has not been published, a suggestion is made to record the change of film thickness with the aid of a photometer. Air or hydrogen bubbles were used (with similar effects), and the equilibrium layers of twice-distilled water on mica

References p. 207

and glass, and films of vaseline oil on steel were investigated. By varying the bubble diameter, or the force which pressed it from above against the base, it was possible to obtain various equilibrium thicknesses and correspondingly a certain section of the isotherm $\Pi = \Pi(h)$. The optical method used for measuring h limited the range of thickness to about 500 Å. Since no theory of thin layers was available at that time, the choice of objects was limited and random, and the results could not be interpreted theoretically.

In 1955 Kusakov and Mekenitskaya[37] used an ingenious conductometric method of checking the thickness of conducting (aqueous solutions) films on glass.

Recently Sheludko and Platikanov (1961)[38] investigated films of benzene on mercury, using the method of automatic interferometric thickness recording over the time of experiment. In studies of the rate of thinning of such layers up to the attaiment of equilibrium thickness, the above authors obtained a very complex $\Pi = \Pi(h)$ isotherm (Fig. 59). Attempts to ap-

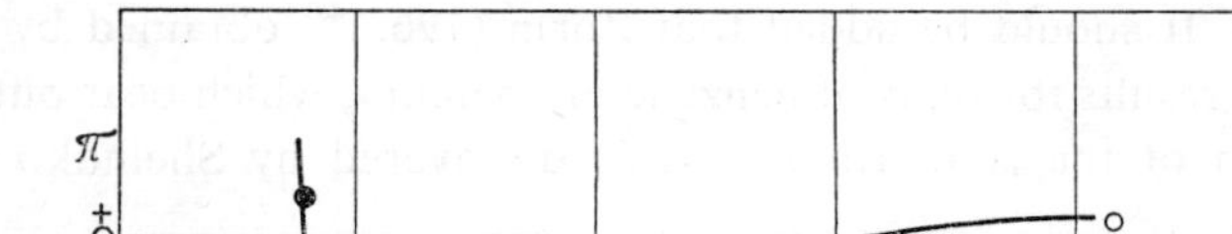

Fig. 59. Isotherm of the disjoining pressure $\Pi(h)$ for a benzene layer on the surface of mercury, obtained from the rate of thinning down to the equilibrium thickness h_r = 240 Å (black dot).

ply this method to the investigation of thin liquid layers on solid bases, *i.e.* to continue the investigation started by Deryagin and Kusakov, were met with exceptional difficulties related mostly to the non-reproducility and inhomogeneity of solid surfaces; this is the cause of the lag occurring in this highly important field.

The only quantitative results obtained so far for thin layers on solid surfaces (optical glasses) appear to be those of Deryagin and Zorin (1955)[39]. This elegant work was highly original in many respects. The excess of the chemical potential in a thin layer was caused by the temperature difference, *i.e.* the base was kept at a slightly lower temperature than the bulk of the liquid, so that the vapour pressure above the base was slightly increased. According to (6.4) this corresponds to a certain drop of $\ln p_h/p_\infty$. The thickness of the polymolecular adsorption layer formed was measured by an ingenious method of elliptical polarisation of the light reflected from the film. Deryagin and Zorin's method is found to be effective when $\Delta\mu$ is relatively large, *i.e.* for small film thicknesses. The interesting results obtained will probably underlie future theoretical interpretation of this most complex case. It should be added that Zorin (1963)[40] obtained by this method some results for films of benzene on mercury, which bear out the complex form of the isotherm $\Pi = \Pi(h)$ discovered by Sheludko and Platikanov[38].

In conclusion, it may be noted that thin layers are an excellent model for many colloidal systems and for the phenomena taking place in them. Thus for instance, free liquid films are a model for foams and the thinning and rupture of free films are a model for foam syneresis and destruction. Thin layers on solid bases reproduce directly the phenomenon of polymolecular adsorption, and their rupture is probably, as already mentioned, a model of the most important stage of the adhesion of particles to bubbles in flotation. Thin liquid layers between condensed media reproduce the situation which arises in the processes of coagulation and coalescence, which is of utmost importance for the theory of these processes. The attempts to investigate these cases [the first of which are due to Deryagin and Obukhov (1935)[1]] will be described in the next chapter in view of their specific characteristics. It should also be pointed out that the investigations of thin layers offer new possibilities for the study of intermolecular forces, and are therefore of great interest in molecular physics.

References p. 207

REFERENCES

1 Deryagin and Obukhov, *Kolloidn. Zh.*, 1 (1935) 385.
2 Frumkin, *J. Phys. Chem.* (*USSR*), 12 (1938) 1.
3 Hamaker, *Physica*, 4 (1937) 1058.
4 Kruyt (Editor), *Colloid Science*, Elsevier, Amsterdam, 1952, Vol. 1.
5 Casimir and Polder, *Phys. Rev.*, 73 (1948) 360; *Nature*, 158 (1946) 787.
6 Keesom, *Phys. Z.*, 22 (1921) 129, 643.
7 Debye, *Phys. Z.*, 21 (1920) 178; 22 (1921) 302.
8 Frenkel', *Kinetic Theory of Liquids*, Dover, New York, 1955, Chapter VI, par. 3 and 4.
9 Deryagin, Abrikosova and Lifshits, *Uspekhi Fiz. Nauk*, 64 (1958) 493.
10 Deryagin, *Bull. Acad. Sci. URSS, Classe Sci., Math., Nat., Sér. Chim.*, (1937) 1153.
11 Bergmann, Löw-Beer and Zocher, *Z. Physik. Chem.*, A181 (1938) 301.
12 Deryagin and Landau, *Zh. Eksperim. i Teor. Fiz.*, 11 (1941) 802.
13 Verwey and Overbeek, *Theory of the Stability of Lyophobic Colloids*, Elsevier, Amsterdam, 1948; Deryagin, *Discussions Faraday Soc.*, No. 18 (1954) 85.
14 Deryagin and Levich, *Dokl. Akad. Nauk, SSSR*, 98 (1954) 985.
15 Deryagin and Martynov, *Kolloidn. Zh.*, 24 (1962) 480.
16 Lifshits, *Zh. Eksperim. i Teor. Fiz.*, 29 (1954) 94.
17 Dzyaloshinskii, Lifshits and Pitaevskii, *Zh. Eksperim. i Teor. Fiz.*, 37 (1959) 229.
18 Rutov, *Teoriya Elektricheskikh Fluktuasii i Teplovogo*, (*Theory of electrical fluctuations and thermal radiation*), Izlucheniya, Moskva, 1953.
19 Abrikosova and Deryagin, *Dokl. Akad. Nauk, SSSR*, 90 (1953) 1055; *Discussions Faraday Soc.*, (1954) 184.
20 Kitchener and Prosser, *Proc. Roy. Soc.* (*London*), A242 (1957) 403.
21 Sparnaay, *Proc. 3rd Intern. Congress of Surface Activity, Cologne, Germany*, (1960) 141.
22 Wells, *Ann. Phys.* (*Paris*), 16 (1921) 69.
23 Deryagin and Titievskaya, *Kolloidn. Zh.*, 15 (1953) 416; *Proc. 2nd Intern. Congress of Surface Activity*, (1957) 211.
24 Abrikosov, Gor'kov and Dzyaloshinkii, *Methods of Quantum Field Theory in Statistical Physics*, Prentice Hall, Englewood Cliffs, 1963.
25 Sheludko, *Proc. Koninkl. Ned. Akad. Wetenschap.*, B65 (1962) 76, 108.
26 Sheludko and Ekserova, *Kolloidn. Zh.*, 165 (1959) 148; 168 (1960) 27.
27 Sheludko, Dessimirov and Nikolov, *Annuaire Univ. Sofia*, 49 (1954/55) 127.
28 Sheludko, *Kolloidn. Zh.*, 155 (1957) 39.
29 Levich, *Physico-chemical hydrodynamics*, Prentice-Hall, Englewood Cliffs, 1962.
30 Frankel and Mysels, *J. Phys. Chem.*, 66 (1962) 190.
31 Sheludko, *Dokl. Akad. Nauk, SSSR*, 123 (1958) 1074.
32 Duyvis, Thesis, Utrecht, 1962.
33 De Vries, *Foam Stability*, Rubberstichting, Delft, 1957, Comm. 326.
34 Sheludko, *Kolloidn. Zh.*, 191 (1963) 52.
35 Deryagin and Kusakov, *Bull. Acad. Sci., URSS, Classe Sci., Math., Nat., Sér. Chim.*, (1937) 1150; *Acta Physicochim., URSS*, 10 (1939) 25, 153.
36 Kusakov, Thesis, 1941, unpublished.
37 Kusakov and Mekenitskaya, *Dokl. IV Mezhdunarodnogo Kongressa v Rime* (*Proceedings of the fourth International Petr. Congress at Rome*), 1955, p. 593.
38 Sheludko and Platikanov, *Kolloidn. Zh.*, 175 (1961) 155.
39 Deryagin and Zorin, *Zhur. Fiz. Khim.*, 29 (1955) 1010.
40 Zorin, *Kolloidn. Zh.*, 25 (1963) 624.

CHAPTER VII

Stability of Lyophobic Sols

The stability of any system, including a colloidal system, is understood to mean its ability to maintain its state and in particular full homogeneity throughout its volume. Unlike molecularly dispersed systems, lyophobic sols have only a limited stability.

Coarse suspensions are unstable mainly because their particles settle at an appreciable rate owing to the force of gravity. In systems with higher degrees of dispersion, in which Brownian movement is sufficiently vigorous to prevent sedimentation, the stability can be disturbed as a result of changes which occur in such systems with time and which lead to an increase in the particle size. When the particle size becomes sufficiently large, the disperse phase separates from the dispersion medium by sedimentation. The particle size may be increased owing to a number of causes.

It is known that fine crystals or droplets are characterised by higher vapour pressure, with correspondingly enhanced solubility. The relationship between linear dimensions of particles and the increase of the saturated vapour pressure or of the solubility is given by the Gibbs-Thomson equation (4.17). It follows from this equation that the effect is noticeable even for particles of colloidal dimensions. In a sufficiently highly disperse heterogeneous system, the larger particles will therefore grow at the expense of the smaller ones. The rate of this process is determined by the rate of diffusion of the solute from particle to particle, and such particle growth develops in the sols of sufficiently soluble substances. It is for instance known that in view of their significant solubility in water, AgCl and $BaSO_4$ form sols which are not particularly stable, and that the stability of the $BaSO_4$ sol can be increased if the solubility of $BaSO_4$ is reduced by the addition of say, alcohol. Processes of recrystallisation-type particle growth are of great importance in gravimetric analysis and in many other applications. Thus in the preparation

References p. 240

of photographic materials, the process of "ripening" of the silver halide emulsions is attributed to recrystallisation-type grain growth*.

For very sparingly soluble substances the velocity of diffusive transfer is insignificant owing to the small number of solute molecules participating in the diffusion process; for this reason the adhesion of colliding particles is in general very much more important in colloidal solutions. Such a coarsening process is called coagulation or flocculation. When the disperse phase of the sol is a liquid or a gas (emulsions and foams), the process can go even further, up to the merging of droplets and bubbles with each other, *i.e.* up to coalescence. Any aggregation of particles as a result of coagulation produces a change in the state of the colloidal system, and even from this point of view is equivalent to the disturbance of its stability. Hence, following Peskov, one can speak of the "aggregative stability" of colloidal systems, which denotes the absence of coagulation. When as a result of coagulation the aggregation proceeds to a sufficiently great extent the system loses its stability and a precipitate is formed owing to the force of gravity.

Aggregation of two particles can only take place when the particles collide. The number of collisions among colloidal particles is therefore of fundamental importance for the coagulation rate. Every collision, however, does not have to be effective, *i.e.* not every encounter results necessarily in aggregation. For the same collision frequency, the effectiveness is determined primarily by the properties of particle surfaces. When adsorbing substances are introduced into the solution these properties can change quite considerably, even in the presence of very small quantities of the added substance, which in turn can affect the effectiveness of the collisions. If this effectiveness is increased, a state can be attained in which all or almost all collisions will result in aggregation. In this case, the rate of coagulation is determined only by the collision frequency and ceases to depend on surface properties; no further change occurs, therefore, on the addition of substances which increase the effectiveness of collisions. This is known as fast coagulation. When on the other hand, not all collisions are effective, the process is termed slow coagulation, and its rate is determined both by the number of collisions and by their effectiveness. This second case is obviously more complex.

Both attraction and repulsion forces participate in the collision between

* It was, however, recently supposed that the coarsening results from coagulation.

a pair of particles. It is generally supposed that the attraction forces are of the van der Waals type, whereas the repulsion forces are due to electrostatic interaction between the electrically charged surfaces of the particles. When the former are predominant, the effectivenees of collisions is high, and the colloidal systems are unstable. When repulsion forces predominate, the effectiveness of colloisions is decreased and the stability is correspondingly higher. It follows that the electrical properties of interfaces, which manifest themselves particularly clearly in electrokinetic phenomena, should play an essential role in the stability of hydrosols. It was found in Chapter V that these properties are very sensitive to the presence of electrolytes.

Coagulation by electrolytes, which as will be shown later is the best understood at the present time, is not, however, the only possibility. Phenomena of the stability of sols in non-polar solvents, *e.g.* of metallic organosols, paints and lacquers, and of many emulsions of water in oil, still remain unexplained. The causes of the stability of such systems, including the mechanism of the action of stabilisers (various soap-like substances—detergents in the case of emulsions) have not as yet been sufficiently studied. Since these causes may also operate in aqueous suspensions, they similarly complicate the mechanism of the stability of such systems. Further discussion will be concentrated almost entirely on the stability of hydrosols, and particularly on the stability factors associated with the particle charge.

Resistance to coalescence, which is of fundamental importance in concentrated emulsions and foams, will be analysed separately in the following chapter.

1. Coagulation by Electrolytes

The sensitivity of colloidal solutions to electrolytes was known even to the pioneer workers such as Selmi, Graham and Faraday. In 1900, Hardy[1] first proposed that the stability of colloids was due to the existence of electrostatic repulsion between colloidal particles. This assumption led to the search for a relationship between the values of the ζ-potential and the stability of colloids. It became clear [*e.g.* Powis (1914)[2]] that when the ζ-potential decreased due to increasing electrolyte concentration, the coagulation rate does in fact increase and that fast coagulation begins at a certain, sufficiently low, value of ζ, known as the critical potential. It is

References p. 240

interesting that in most cases the value of the critical potential does not depend on the type of electrolyte used to attain it. Thus *e.g.* for the As_2S_3 sols (Powis) and Fe_2O_3 sols (Gosh) fast coagulation after the addition of various electrolytes is attained at almost the same value of ζ, although the corresponding electrolyte concentrations are of course completely different from each other (Table VII, 1).

TABLE VII, 1

CRITICAL CONCENTRATION AND CRITICAL ζ-POTENTIAL

As_2S_3 *sol* (—)			Fe_2O_3 *sol* (+)		
Electrolyte	*Critical concentration* m equiv./l	*Critical value of* ζ (mV)	*Electrolyte*	*Critical concentration* m equiv./l	*Critical value of* ζ (mV)
KCl	40.0	44	KCl	100.0	33.7
$BaCl_2$	1.0	26	NaOH	7.5	31.5
$AlCl_3$	0.15	25	K_2SO_4	6.6	32.5
$Th(NO_3)_4$	0.20	27	Aniline-sulphate	8.0	31.4
	0.28	26			
	0.40	24	K_2CrO_4	6.5	32.5
			$K_3Fe(CN)_6$	0.65	30.2

In 1882 Schulze[3] showed that various electrolytes display different powers of flocculating the As_2S_3 sol and that in this case the decisive role is played by cations, particularly by polyvalent cations. Hardy (1900)[1] confirmed these results and generalised them as follows: the coagulating ion of an electrolyte is the one charged oppositely to the colloidal particles. This conclusion is confirmed by Table VII,2, which contains results obtained by Freundlich[4] for the As_2S_3 sol, and by Lepin and Bromberg (1939)[5] for the Fe_2O_3 and Al_2O_3 sols.

The flocculating power of an electrolyte is characterised by the minimum concentration c_{cr} at which fast coagulation is attained. Since the transition from slow to fast coagulation takes place gradually with increasing electrolyte concentration, the determination of c_{cr} is not very sharp. Values of c_{cr} obtained by various authors or by the same authors but by different methods should not therefore be regarded as strictly comparable.

The data given in Tables VII, 1 and 2 show that the negatively charged

TABLE VII, 2

c_{cr}-VALUES OF ELECTROLYTES

Electrolyte	m equiv./l	*Electrolyte*	m equiv./l
(*a*) As_2S_3 *sol (negatively charged)*			
CH_3COOK	110.0	$MgCl_2$	0.717
HCOOK	86.0	$MgSO_4$	0.810
LiCl	58.4	$CaCl_2$	0.649
NaCl	51.0	$SrCl_2$	0.635
KNO_3	50.0	$BaCl_2$	0.691
KCl	49.5	$Ba(NO_3)_2$	0.687
K_2SO_4	65.6	$ZnCl_2$	0.685
NH_4Cl	42.3	$UO_2(NO_3)_2$	0.642
HCl	30.8	Quinine sulphate	0.240
H_2SO_4	30.1	$AlCl_3$	0.093
Guanidine nitrate	16.4	$Al(NO_3)_3$	0.095
Strychnine nitrate	8.0	$Al_2(SO_4)_3$	0.096
Aniline hydrochloride	2.52	$Ce(NO_3)_3$	0.080
Morphine hydrochloride	0.425	$Ce_2(SO_4)_3$	0.092
Crystal violet	0.165		
Fuchsine NB	0.114		
(*b*) AgI *sol (negatively charged)*			
KCl	255.5	$CaSO_4$	10.00
KNO_3	260.0	$MgSO_4$	11.32
K_2CrO_4	272.0	$Al(NO_3)_3$	0.0616
K_2SO_4	272.0	$La(NO_3)_3$	0.0750
$K_3Fe(CN)_6$	284.0	$Co(NH_3)_6(NO_3)_3$	0.0690
$K_4Fe(CN)_6$	400.0	$Co(NH_3)_6Cl$	0.0660
$Ba(NO_3)_2$	6.00	$Al_2(SO_4)_3$	0.150
$Sr(NO_3)_2$	7.00	$Ce_2(SO_4)_3$	0.138
$BeSO_4$	6.00	$Th(NO_3)_4$	0.028

sols (As_2S_3) are particularly sensitive to cations, and that cations with the same valency are more or less equally active (similar values of c_{cr}), whilst the positive sols (Fe_2O_3, Al_2O_3) are sensitive to anions. The activity of inorganic ions is the higher (*i.e.* the values of c_{cr} are lower) the higher their valency. The activities of divalent cations are roughly 80 times greater and that of the trivalent ions 640 times greater than those of the alkali ions (Schulze–Hardy rule). Organic ions characterised by a high adsorption potential are particularly active.

References p. 240

TABLE VII, 2 (continued)

Electrolyte	Fe_2O_3 *sol* (in m equiv./l)	Al_2O_3 *sol* (in m equiv./l)
(c) Fe_2O_3 *and* Al_2O_3 *sols (positively charged)*		
NaCl	9.25	77
KCl	9.0	80
NH_4Cl	—	100
$BaCl_2$	9.65	—
KBr	12.5	150
KI	16.0	300
KCNS	—	112
KNO_3	12.0	—
$Ba(NO_3)_2$	14.0	—
HCl	400.0	—
$Ba(OH)_2$	0.42	—
Potassium benzoate	—	13
Potassium salicylate	—	8
Potassium picrate	—	4
K_2SO_4	0.205	0.28
Tl_2SO_4	0.22	—
$MgSO_4$	0.22	—
$K_2Cr_2O_7$	0.195	—
H_2SO_4	0.5	—
$K_3Fe(CN)_6$	—	0.10
K_2CrO_4	—	0.60
Potassium oxalate	—	0.36
Potassium citrate	—	0.30
$K_4Fe(CN)_6$	—	0.08

Although ions of the same valency display close values of activity, their effect is not identical, and they can be arranged in a lyotropic series of ascending flocculating power, *i.e.*

$$Li^+ < Na^+ < K^+ < NH_4^+ ;$$

in other words c_{cr} decreases from Li^+ to NH_4^+ [Lepin, Bromberg (1939)[5]].

Freundlich (1926)[6] showed that the flocculating power of organic cations increases with increasing chain length, corresponding to the increasing adsorption power in accordance with Traube's rule.

These observations, which by no means exhaust the great variety of phenomena occurring during flocculation by electrolytes, point to the fact that it is the state of the interphase surface, and particularly its electric properties, that are an important factor in the stability of lyophobic colloids.

Observations supporting the assumption of the particularly strong influence of the electric state of the interface on the stability of lyophobic colloids can be briefly summarised as follows:

(1) Coagulation is caused by the ion carrying a charge opposite to that of the colloid particle.

(2) The rates of slow coagulation are similar at similar values of the ζ-potential.

(3) The coagulating power of the ions increases with their valency, the increase of which (according to the theory of the electrical double layers) results, at equal concentrations, in a greater lowering of the ζ-potential.

(4) The strongly adsorbed organic ions, which according to Stern's theory cause a particularly effective lowering of the ζ-potential, possess a particularly high coagulating power.

The parallel effects of electrolytes on the stability of hydrosols and on the ζ-potential of the disperse phase/solvent interface are shown particularly clearly by the polyvalent and organic ions which can reverse the charge of the interface. In such cases a sudden lowering of the stability of the colloid may be observed as the electrolyte concentration is increased, followed by fast flocculation. A second zone of stability is, however, attained at high electrolyte concentrations. This phenomena can easily be explained by the acquisition of an opposite charge by the colloid. Indeed, at certain electrolyte concentrations the colloid acquires a charge (now opposite in sign to the previous one) which ensures its stability. Of course, in every case the ζ-potential drops to zero at sufficiently high concentrations of the electrolyte, and the colloid simultaneously becomes unstable.

The phenomena considered above bear out qualitatively the decisive influence of the electric properties of the particle surface on the stability of hydrosols.

Although the quantitative picture has not as yet been fully developed, it will be shown that in some cases it is already possible to carry out a quantitative analysis of such phenomena.

It should be added that, besides the facts which point to the decisive effect of the ζ-potential on the stability of hydrosols, observations are available which indicate that the stability is also influenced by other factors. Thus, for example, Kargin (1935, 1938)[7] demonstrated that highly

References p. 240

dialysed sols of SiO_2, Al_2O_3 etc. preserve complete stability even when ζ approaches zero (as established by the electrophoretic method).

2. The Kinetics of Fast Coagulation

The present section is concerned with the rate of coagulation, which is measured by the decrease in the concentration of the colloidal particles in unit time. Since this rate usually varies with time, it is expressed as the derivative of concentration with respect to time (the differential rate).

It is perhaps convenient to begin with a study of the fast coagulation as this is the simpler case, and then to extend the results to the case of slow coagulation. The problem will then consist of two parts: In the first place it is necessary to determine the number of collisions taking place between the particles, and secondly, it is necessary to explain the nature of these collisions, *i.e.* the problem of collisions resulting in conglomeration, and of ineffective collisions, when the colliding particles again move away from each other.

Let us analyse the case when, at a certain moment $t = 0$, conditions are created in a monodisperse sol which lead to particle aggregation after every collision, *e.g.* when a sufficient quantity of electrolyte is added to the colloidal solution. Double particles are formed first, which in turn collide with single particles to form triple aggregates, with other double particles to form quadruple aggregates, etc. The problem, solved by Smoluchowski[8] in 1918, consisted in finding the change of the number of particles in time, regardless of their size.

Let us first analyse the formation of the double particles. This case is completely analogous to a bimolecular reaction, and its rate is given by the corresponding rate equation:

$$\frac{dc_1}{dt} = -kc_1c_2, \tag{7.1}$$

where c_1 and c_2 are concentrations of the reactants 1 and 2, t is the time and k is the velocity constant. In the present case both particles are identical, so that

$$\frac{dc}{dt} = -kc^2, \tag{7.2}$$

where c is the concentration of the original single particles.

The integration of (7.2) between the limits from c_0 at $t = 0$ to a certain concentration c at $t > 0$ yields

$$c = \frac{c_0}{1 + kc_0 t}. \tag{7.3}$$

The product $kc_0 t$ is a dimensionless magnitude and kc_0 has therefore the dimensions of $[\text{sec.}]^{-1}$. The time T at which concentration c will be half of c_0 is, according to (7.3)

$$T = \frac{1}{kc_0}. \tag{7.4}$$

The product kc_0 is thus a reciprocal of the time T during which the coagulation is semi-complete, and the expression (7.3) can be written as

$$c = \frac{c_0}{1 + \dfrac{t}{T}}. \tag{7.5}$$

According to the accepted ideas, the rate of coagulation is determined by the frequency of collisions between particles. This frequency depends on the concentration of particles and on the intensity of their Brownian movement, which in turn is characterised by the diffusion constant. Taking this into account, let us calculate the constant T on the basis of the diffusion mechanism by which the particles approach each other. Let us first solve the problem for one fixed particle. The collision with this particle (radius r) occurs whenever any other particle approaches it so closely that the distance between the centres of both particles, if they are spherical, is equal to $2r$. In other words, we are considering the diffusion of point-masses towards a sphere of radius $R = 2r$. At the same time, the concentration of single particles on the surface of this sphere is equal to zero, since here the single particles cease to exist as such. We therefore obtain the following boundary conditions:

$$\begin{aligned} &\text{at } t = 0\,, \quad c = c_0 \quad \text{for } \rho > R \\ &\text{at } t > 0\,, \quad c = 0 \quad \text{for } \rho = R\,, \end{aligned} \tag{7.6}$$

where ρ is the distance from the centre of a sphere of radius R, and $t = 0$ is the initial moment of diffusion. Fick's second law of bulk diffusion is given by

References p. 240

$$\frac{\partial c}{\partial t} = D\Delta c, \quad \Delta c = \frac{\partial^2 c}{\partial x^2} + \frac{\partial^2 c}{\partial y^2} + \frac{\partial^2 c}{\partial z^2}.$$

In the considered case of spherical symmetry, it is more convenient to write Fick's second law as

$$\frac{\partial c}{\partial t} = D\left(\frac{\partial^2 c}{\partial \rho^2} + \frac{2}{\rho}\frac{\partial c}{\partial \rho}\right), \tag{7.7}$$

where the distance ρ from the centre of symmetry is substituted for the coordinates x, y, z. The solution of eqn. (7.7) has the form

$$c = c_0\left\{1 - \frac{R}{\rho}\left(1 - \frac{2}{\sqrt{\pi}}\int_0^{\frac{\rho - R}{2\sqrt{Dt}}} e^{-\xi^2}\,d\xi\right)\right\}. \tag{7.8}$$

In the stationary state* [when $t \to \infty$ and the integral in (7.8) tends to zero] the concentration at a distance ρ does not depend on time t, so that

$$c = c_0\left(1 - \frac{R}{\rho}\right). \tag{7.8'}$$

The flux of particles towards the sphere, *i.e.* the number of particles which cross the surface of a sphere of radius R in 1 sec is, according to Fick's first law,

$$I = 4\pi R^2 D\left(\frac{\partial c}{\partial \rho}\right)_{\rho=R}. \tag{7.9}$$

and has the same magnitude for all other values of ρ.

The derivate $(\partial c/\partial \rho)_{\rho=R}$ is calculated from (7.8'):

$$\left(\frac{\partial c}{\partial \rho}\right)_{\rho=R} = \frac{c_0}{R} \tag{7.10}$$

whence

$$I = 4\pi R D c_0. \tag{7.11}$$

The case analysed differs from actual practice in that the particle with respect to which the diffusion is investigated is not in fact fixed, but is endowed with the same mobility as the particles which diffuse towards it. It is therefore necessary to analyse not a movement of diffusing par-

* It can be proved that this state is established almost at once.

ticles towards a fixed particle, but the relative motion of two moving particles. If the mean-square displacement of one particle is $\overline{x_1^2}$ and that of the other is $\overline{x_2^2}$, during the same time t, then, after introducing the corresponding diffusion coefficients D_1 and D_2 we obtain from Einstein's formula for the theory of Brownian movement (see Chapter II):

$$D_1 = \frac{\overline{x_1^2}}{2t} \quad \text{and} \quad D_2 = \frac{\overline{x_2^2}}{2t}.$$

On the other hand, the mean-square displacement of one particle with respect to another during the same time is

$$\overline{(x_1 - x_2)^2} = \overline{x_1^2} - 2\overline{x_1 x_2} + \overline{x_2^2} = D_1 2t + D_2 2t = (D_1 + D_2)2t \tag{7.12}$$

since the mean displacements $\overline{x_1}$ and $\overline{x_2}$ are equal to zero*.

In the case of two identical particles we have

$$\overline{(x_1 - x_2)^2} = 4D_1 t = 4D_2 t \quad \text{or} \quad D = \frac{\overline{(x_1 - x_2)^2}}{2t} = 2D_1.$$

For the flux I of single particles towards the sphere of radius R, which moves along with one of the particles we have therefore in place of (7.11):

$$I = 8\pi R D_1 c_0. \tag{7.13}$$

The flux I expresses the rate of aggregation with respect to one particle, and will be c_0 times greater with respect to all c_0 particles in 1 cm^3, so that

$$\frac{dc}{dt} = -\, 8\pi R D_1 c^2. \tag{7.14}$$

This equation is valid for any concentration, and for this reason c has been substituted for c_0. It must be taken into consideration that during each collision two primary particles disappear, but at the same time the collision frequency is doubled since each collision is taken into account twice: first, when one of both colliding particles is regarded as the central particle, and again when the second particle is regarded as the central particle.

* It may be proved in statistics that if x_1 and x_2 are independent, as in the present case, then $\overline{x_1 x_2} = \overline{x_1} \cdot \overline{x_2}$.

References p. 240

By comparing (7.14) with (7.2) we obtain the expression for the rate constant of fast coagulation

$$k = 8\pi R D_1, \tag{7.15}$$

and for the time T, during which coagulation is semi-complete, according to (7.4)

$$T = \frac{1}{8\pi R D_1 c_0}. \tag{7.16}$$

Aggregation of more complex particles occurs by an analogous scheme. Having determined from eqn. (7.14) the number of primary particles which had disappeared during 1 sec, we then find the number of collisions taking place during 1 sec in 1 cm^3. Since in every collision two primary particles take part, this number is clearly

$$z_{11} = 4\pi D_1 R c^2 = 2\pi D R c_1^2. \tag{7.17}$$

The collision frequency between particles of categories i and j (in 1 cm^3)*

$$z_{ij} = 4\pi D_{ij} R_{ij} c_i c_j \tag{7.18}$$

where $D_{ij} = D_i + D_j$, $R_{ij} = r_i + r_j$, and r_i and r_j are the effective radii of the aggregates of categories i and j. When i and j particles collide, particles of category k are formed, where $k = i + j$. These k-particles are also formed during collisions of particles of categories $i-1$ and $j+1$, $i-2$ and $j+2$ etc., up to $i=1$ and $j=k-1$. Thus in a system which contains particles of any category, the number of particles of category k, which are formed during collisions taking place in 1 sec in 1 cm^3 is

$$\frac{1}{2} \sum_{\substack{i+j=k \\ i=1,2\ldots \\ j=1,2\ldots}} 4\pi D_{ij} R_{ij} c_i c_j .$$

The factor of $\frac{1}{2}$ is introduced to allow for the fact that each term is counted twice during the summation.

At the same time, all particles of category k, are transformed into particles of a higher category when meeting any other particle. The number

* By substituting $c_i c_j$ for c_1^2, double counting of the collisions is avoided.

of k-particles decreases therefore during such collisions by

$$\sum_{i=1}^{i=\infty} 4\pi D_{ik} R_{ik} c_i c_k .$$

As a result of the formation and disappearance of k-particles, their increment during 1 sec in 1 cm^3 amounts to

$$\frac{dc_k}{dt} = \tfrac{1}{2} \sum_{\substack{i+j=k \\ i=1,2\ldots \\ j=1,2\ldots}} 4\pi D_{ij} R_{ij} c_i c_j - \sum_{i=1}^{i=\infty} 4\pi D_{ik} R_{ik} c_i c_k . \tag{7.19}$$

Smoluchowski simplified this expression by assuming that

$$D_{ij} R_{ij} = 2 D_1 R_1 , \tag{7.20}$$

where D_1 and R_1 are the diffusion coefficient and the diameter of primary particles, respectively. Application of (7.20), which is possible because the diffusion coefficient vary inversely with the radii of the diffusing particles, is fully justified only in cases when the radii of both particles are similar to each other. More accurate calculations for a polydisperse sol are given by Müller (1926)[9].

With the aid of the approximation (7.20), expression (7.19) may be simplified as follows:

$$\frac{dc_k}{dt} = 4\pi D_1 R_1 \{ \sum_{\substack{i+j=k \\ i=1,2\ldots \\ j=1,2\ldots}} c_i c_j - 2 \sum_{i=1}^{i=\infty} c_i c_k \} . \tag{7.21}$$

After carrying out summation with respect to all k, we obtain the rate of decrease of the total number of particles as

$$\frac{d \sum_{k=1}^{k=\infty} c_k}{dt} = 4\pi D_1 R_1 \{ \sum_{k=1}^{k=\infty} \sum_{\substack{i+j=k \\ i=1,2\ldots \\ j=1,2\ldots}} c_i c_j - 2 \sum_{k=1}^{k=\infty} \sum_{i=1}^{i=\infty} c_i c_k \} .$$

It is easy to show (*e.g.* graphically) that

$$\sum_{k=1}^{k=\infty} \sum_{\substack{i+j=k \\ i=1,2\ldots \\ j=1,2\ldots}} c_i c_j = \sum_{k=1}^{k=\infty} \sum_{i=1}^{i=\infty} c_i c_k .$$

References p. 240

Besides, since both summations on the right hand side have the same limits,

$$\sum_{k=1}^{k=\infty} \sum_{i=1}^{i=\infty} c_i c_k = (\sum_{k=1}^{k=\infty} c_k)^2 ,$$

so that

$$\frac{\mathrm{d} \sum_{k=1}^{k=\infty} c_k}{\mathrm{d}t} = - 4\pi D_1 R_1 (\sum_{k=1}^{k=\infty} c_k)^2 . \tag{7.22}$$

This expression is analogous to the rate equation for a bimolecular reaction since $\sum_{k=1}^{k=\infty} c_k$ is the total number of particles in 1 cm^3. Integration of (7.22) gives

$$\sum_{k=1}^{k=\infty} c_k = \frac{c_0}{1 + 4\pi D_1 R_1 c_0 t} = \frac{c_0}{1 + \frac{t}{T}} \tag{7.23}$$

since the original number of particles is c_0 at $t = 0$.

Making use of (7.23) and of the differential equation (7.21) it is possible to calculate the number of particles of each category as a function of time. Let us first obtain the rate of the decrease of concentration of single particles ($k = 1$).

At $k = 1$, the sum in (7.21)

$$\sum_{\substack{i+j=k \\ i=1,2\ldots \\ j=1,2\ldots}} c_i c_j$$

is equal to zero, since single particles are not formed as a result of collisions; $c_k = c_1$ and

$$\sum_{i=1}^{i=\infty} c_i = \frac{c_0}{1 + \frac{t}{T}}$$

according to (7.23). Moreover, according to (7.16), $4\pi DR = 1/c_0 T$, where $D = 2D_1$. Then

$$\frac{\mathrm{d}c_1}{\mathrm{d}t} = - \frac{2c_1}{c_0 T} \frac{c_0}{1 + \frac{t}{T}} \quad \text{or} \quad \frac{\mathrm{d}c_1}{c_1} = - 2 \frac{\mathrm{d}\left(\frac{t}{T}\right)}{1 + \frac{t}{T}} .$$

After integration, raking into account the initial conditions ($t = 0$, $c_1 = c_0$) we obtain

$$\ln \frac{c_1}{c_0} = 2 \ln \left(1 + \frac{t}{T} \right), \text{ or}$$

$$c_1 = \frac{c_0}{\left(1 + \frac{t}{T} \right)^2}. \tag{7.24}$$

Similarly for c_2

$$\frac{dc_2}{dt} = \frac{1}{c_0 T} \left\{ c_1^2 - 2c_2 \frac{c_0}{1 + \frac{t}{T}} \right\},$$

or, after substituting c_1 from (7.24), we obtain

$$\frac{dc_2}{dt} = \frac{1}{c_0 T} \left\{ \frac{c_0^2}{\left(1 + \frac{t}{T} \right)^4} - 2 c_2 \frac{c_0}{1 + \frac{t}{T}} \right\}.$$

This differential equation is solved according to the scheme

$$\frac{dy}{dx} = A(x)y + B(x), \quad y = e^{\int A\,dx} \left(h + \int B(x) e^{-\int A\,dx} dx \right),$$

where A and B are functions of x, and h is an integration constant. In the present case:

$$\frac{dc_2}{dt} = -\frac{2c_2}{T\left(1 + \frac{t}{T} \right)} + \frac{c_0}{T\left(1 + \frac{t}{T} \right)^4}.$$

Assuming that $x = t$ and $y = c_2$, we obtain

$$A(t) = -\frac{2}{T\left(1 + \frac{t}{T} \right)} \quad \text{and} \quad B(t) = \frac{c_0}{T\left(1 + \frac{t}{T} \right)^4};$$

$$\int A\,dt = -2 \int \frac{d(T+t)}{T+t} = -2 \ln (T+t) = \ln (T+t)^{-2};$$

$$\int B e^{-\int A\,dt} = \int \frac{c_0 (T+t)^2}{T\left(1 + \frac{t}{T} \right)^4} dt = -c_0 T^3 \frac{1}{T+t}.$$

References p. 240

Then

$$c_2 = \frac{h - c_0 T^3/(T+t)}{(T+t)^2}.$$

The initial conditions for double particles, as indeed for all aggregated particles, are $t = 0$, $c_2 = 0$, and the integration constant is therefore

$$h = c_0 T^2,$$

or, finally,

$$c_2 = c_0 \frac{t/T}{(1 + t/T)^3}. \qquad (7.25)$$

In analogy with (7.24) and (7.25) we obtain for the particles of the k-th category

$$c_k = c_0 \frac{(t/T)^{k-1}}{(1 + t/T)^{k+1}}.$$

Variations in the numbers of particles of various categories with time during coagulation are shown graphically in Fig. 60. It may be seen that the number of single particles decreases monotonically from c_0 at $t = 0$,

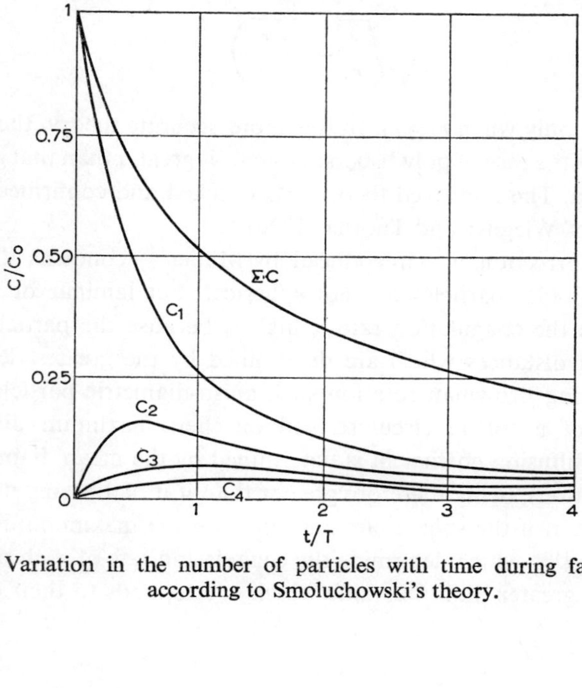

Fig. 60. Variation in the number of particles with time during fast coagulation, according to Smoluchowski's theory.

since larger aggregates are formed at the expense of primary particles. At the same time, the number of higher order particles, which is zero at the beginning, increases at the expense of the lower order particles and reaches a maximum, after which it begins to decrease owing to the formation of particles of a still higher order. The total number of all particles, $\sum c$, decreases monotonically with time, although less steeply than the number of single particles, as a result of the growing number of large aggregates.

It should also be noted that the constant $T = 1/4\pi D_1 R_1 c_0$ does not depend, within the scope of these approximations, on the particle size. As noted before, Smoluchowski's theory of suspensoids which contain particles widely differing in size was supplemented by Müller[9] who developed a more accurate form of condition (7.20). Thus $D_{ij} = D_i + D_j$, $R_{ij} = r_i + r_j$ and, finally, $D \sim 1/r$ so that, more accurately,

$$D_{ij}R_{ij} = D_1 r_1 \left(\frac{1}{r_i} + \frac{1}{r_j}\right)(r_i + r_j) =$$

$$= D_1 r_1 \left\{4 + \left(\sqrt{\frac{r_i}{r_j}} - \sqrt{\frac{r_j}{r_i}}\right)^2\right\} > 4D_1 r_1 = 2D_1 R_1.$$

The term

$$\left(\sqrt{\frac{r_i}{r_j}} - \sqrt{\frac{r_j}{r_i}}\right)$$

becomes zero only when $r_i = r_j$. In this more accurate theory, the rate of coagulation in the case of polydisperse colloids is greater than that given by Smoluchowski. The improved theory was checked and confirmed by the experiments of Wiegner and Tuorila (1926)[10].

A further correction, also introduced by Müller[11], concerns the cases when the colloidal particles are not spherical, but laminar or rodlike. In such cases the coagulation rate is higher, because the particles meet each other at distances which are determined by the greatest length of particles (during Brownian rotation such aniso-diametric particles "fill" the volume of a sphere circumscribed on their maximum diameter), whereas the diffusion coefficient is determined by the mean diameter (*cf.* Perrin's equation), being therefore greater than if it had been determined by the diameter of the sphere circumscribed on the maximum diameter. The products DR, which determine the coagulation rate of such particles, are therefore greater than the value which corresponds to their effective

References p. 240

volume. Müller's postulate has been confirmed experimentally; thus for rodlike particles Wiegner and Marshall (1929)[12] obtained a rate which was higher by a factor of 50 than the rate calculated without taking the above considerations into account.

Finally, it should be noted that the rate of coagulation increases when the solution is stirred, and in the case of the sedimentation of polydisperse suspensions (orthokinetic coagulation). The theory of these special cases was developed by Smoluchowski[13], Tuorila[14] and Müller[15]. The reason for the increased rate of coagulation is the fact that if another motion, the velocity of which is not uniform for various particles, is superimposed on the Brownian movement (as during both stirring and sedimentation in which mostly the heavier particles take part), the prob-

TABLE VII, 3

EXAMPLES OF FAST COAGULATION

Time t, sec	*The number of particles* (per ml) ($\times 10^{-8}$)		*Time of semi-coagulation T* sec
	Calculated	*Found*	
	Gold sol: $r_1 = 512$ Å, at 279° abs.		
0	20.20	20.20	—
30	14.40	14.70	80
60	11.19	10.80	69
120	7.74	8.25	83
240	4.78	4.89	76
480	2.71	3.03	85
			average 79
	Kaolin suspension, at 330° abs.		
0	5.0	5.0	—
105	3.80	3.90	372
180	3.23	3.18	314
255	2.82	2.92	358
355	2.46	2.52	335
420	2.20	2.00	280
510	1.96	1.92	318
600	1.77	1.75	313
1020	1.22	1.54	(452)
2340	0.62	1.15	(699)

ability of collisions between the particles is increased. Since this effect is determined by the velocity of the relative displacement of particles it is clear that the increase in the rate of coagulation during stirring does not depend on the degree of polydispersity, whilst an increase of the coagulation rate during sedimentation is observed only in polydisperse sols whose particles settle at different rates.

Experimental investigations of fast coagulation yielded a good confirmation of the theory, in spite of the complexity of such experiments. The work of Zsigmondy (1917)[16], who as a matter of fact placed this problem before Smoluchowski, is of great importance in this context. The above studies were mostly carried out with monodisperse gold sols (Zsigmondy, Westgren[17], Kruyt[18], Tuorila and others), in which the number of particles in a certain volume was determined at intervals of time. Table VII, 3 shows the results of measurements carried out by Tuorila in 1926[19], with a gold sol and with a suspension of kaolin.

3. The Kinetics of Slow Coagulation

It has already been pointed out that the difference between fast and slow coagulation consists in the fact that all collisions are effective in the former case, whereas in the latter only a fraction α of the collisions results in aggregation. The coefficient α was introduced by Smoluchowski. It is clear that $\alpha = 1$ in the case of fast coagulation. Expressions derived in the development of the theory of fast coagulation may be applied with certain reservations to the cases when $\alpha < 1$, but the time scale must be correspondingly adjusted by using, instead of (7.16), a more general equation

$$T = \frac{1}{8\pi R_1 D_1 c_0 \alpha}. \qquad (7.26)$$

The relation between α and the physico-chemical properties of the interface, *e.g.* the ζ-potential is of particular theoretical and practical interest, as this would complete the development of the theory of slow coagulation and of the theory of the stability of lyophobic colloids in general: the phenomenon would thus find an ultimate explanation. No complete solution of this fundamental problem has, however, been achieved up to the present time.

A rigorous calculation of the rate of slow coagulation was suggested

References p. 240

in 1934 by Fuchs[20], who replaced the conventional equations for diffusion in the theory of fast coagulation by the more complex expressions for the diffusion in a force field which takes into account the interactions occurring between particles which are very close to one another. The following formula was thus put forward for the diffusion with respect to a fixed particle, in place of (7.9):

$$I = 4\pi R^2 \left(D \frac{\mathrm{d}c}{\mathrm{d}\rho} + \frac{c}{B} \frac{\mathrm{d}V(\rho)}{\mathrm{d}\rho} \right)_{\rho=R} \tag{7.27}$$

where B is the friction constant of the particle ($6\pi\eta r$ for a sphere), and $V(\rho)$ is the potential energy of the interaction between two particles separated by a centre-to-centre distance ρ.

Elaboration of formula (7.27) leads to rather complex mathematics, which will not be considered in this book. It should, however, be noted that in this case the potential energy of interaction between two particles is introduced, which as a matter of fact determines the effectiveness of the collision.

Particles repelled from one another must possess sufficient kinetic energy to overcome this repulsion. Since this energy may vary to a large extent, only a fraction of the colliding particles will aggregate, *i.e.* only those particles whose kinetic energy exceeds the maximum repulsion. The mean kinetic energy of thermal *i.e.* Brownian movement is measured by the product kT, so that the effectiveness of collisions is determined by the ratio between $V_{\max}$ and kT, where $V_{\max}$ is the maximum energy of interaction (repulsion) between the two particles. In other words, the rate of slow coagulation depends on the value of $V_{\max}$ for the particles concerned.

If the interaction energy between particles has a repulsion maximum $V_{\max}$ for a certain distance between their centres then, according to the Boltzmann–Einstein formula, the probability of overcoming this energy barrier is proportional to $\exp(-V_{\max}/kT)$. On the other hand, α, as the ratio of the number of aggregated particles to the number of all particles which underwent collisions, provides a measure of the probability of aggregation (*i.e.* of the overcoming of the barrier $V_{\max}$). It follows therefore that ultimately $\alpha \sim \exp(-V_{\max}/kT)$, as from the theory of Fuchs.

4. Interaction between Particles

If the stability of lyophobic sols is to be explained on the basis of the interaction energy, this energy must exhibit a certain repulsion maximum at some distance ρ, the value of this repulsion being a function of electrolyte concentration, *i.e.* decreasing when the concentration is increased. Furthermore, the repulsion must decrease at small distances and pass into attraction, to allow the aggregation of particles which had overcome the repulsion barrier.

Numerous observations of the favourable effect of particle charge on the stability of the resulting sols indicate directly the nature of the repulsion. The origin of the repulsion forces is obviously to be sought in the diffuse electric layers. It should be noted that the earlier rather naive ideas, according to which the repulsion is due to direct coulombic interactions between similarly charged particles, are definitely incorrect. The colloidal particles of a sol possesses a charge with respect to the aqueous solution but, together with the surrounding diffuse ionic layer of opposite charge, constitutes an electrically neutral entity. Interaction between such particles can only take place owing to the deformation of the surrounding ionic atmospheres when the particles approach closely to each other, and it should be described quantitatively on this basis.

It was originally attempted (Langmuir)[21] to deduce the second (attraction) component on the basis of the same ideas, assuming that at sufficiently small particle separations the ionic atmospheres were overlapping and gave rise to attraction necessary for the aggregation. These attempts proved, however, to be unsuccessful.

In 1937 Hamaker[22] suggested van der Waals' forces as the source of attraction and demonstrated that the overall effect of attraction between all molecules of one particle and all molecules of another decreases with distance much more slowly than the energy of interaction between the individual molecules. This result imparts physical meaning to such a treatment.

Calculation of the interaction energy between the sol particles as a sum of the electrostatic (V_{el}) and van der Waals (V_{vw}) components requires an allowance for particle size and shape. Such calculations, *e.g.* for spherical particles, can be found in the bibliography[23,24]. In the present case only a hypothetical interaction between two identical platelike particles separated by gaps which are very narrow as compared with the linear dimensions

References p. 240

of the particles will be considered. This extreme simplification is possible because we shall confine ourselves to explaining, on its basis, the very strongly expressed effect of the valency of the coagulating ion on its critical concentration. Such a model is of course unreliable, to say the least, for an explanation of the finer effects.

It should be readily appreciated that for such types of gap the problem can be solved by applying the results obtained for relatively thin plane-parallel layers, described in the preceding chapter, where the force of repulsion per 1 cm^2 of the layer, which arises from the deformation of the diffuse electrical layers, was approximately given by (6.17):

$$\Pi_{el} = 64cRT\gamma^2 e^{-\varkappa h}.$$

The potential energy of repulsion can be easily obtained by integrating (6.17) in the form

$$V_{el} = -\int_{\infty}^{h} \Pi_{el} \, dh = \frac{64cRT\gamma^2}{\varkappa} e^{-\varkappa h}, \qquad (7.28)$$

where h is the width of the gap between particles, which is filled with the solution of a binary electrolyte of concentration c, and which contains two identical deformed diffuse electric layers of thickness

$$1/\varkappa = \sqrt{\frac{\varepsilon RT}{8\pi z^2 F^2 c}},$$

[according to (5.33)] in the undeformed state.

The energy of attraction can be calculated just as easily, using the same formula ($V_{vw} = -\int_{\infty}^{h} \Pi_{vw} \, dh$), from the van der Waals component of the force of attraction per 1 cm^2, on the basis of the expression (Chapter VI)

$$\Pi_{vw} = -K/h^3.$$

This formula is also applicable to an empty gap. It can be shown[25] that when the gap is filled with a liquid, the nature of the relation $\Pi_{vw} \sim 1/h^3$ remains unchanged and only the absolute value of K is affected.

Consequently

$$V_{vw} = -\int_{\infty}^{h} \Pi_{vw} \, dh = -\frac{K}{2h^2}, \qquad (7.29)$$

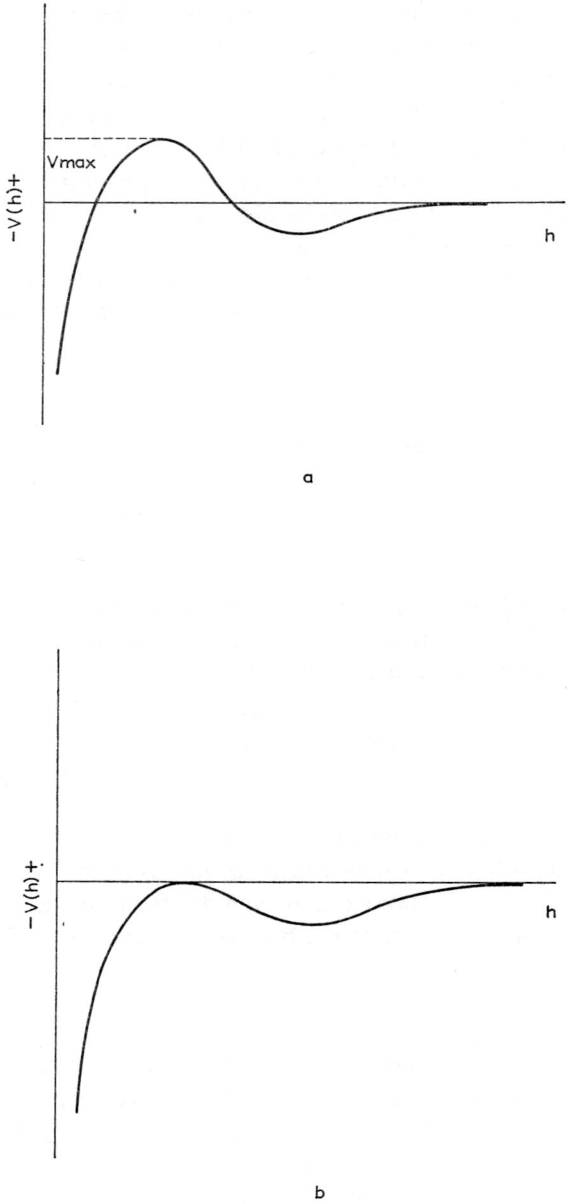

Fig. 61. Dependence of the interaction energy $V(h)$ on the distance h between particles: (a) during slow coagulation; (b) when fast coagulation sets in.

References p. 240

and the total interaction energy is obtained as

$$V = V_{el} + V_{vw} = \frac{64cRT\gamma^2}{\varkappa} e^{-\varkappa h} - \frac{K}{2h^2}. \qquad (7.30)$$

The dependence (7.30), $V = V(h)$, satisfies the requirements of the problem. For a certain ratio of constants it has the shape shown in Fig. 61a. For large values of h, V is negative (attraction), following the energy of attraction V_{vw}, which decreases more slowly with increasing distance ($\sim 1/h^2$). At short distances (small h), the positive component V_{el} (repulsion), which increases exponentially with decreasing h, ($\sim e^{-\varkappa h}$), can overcompensate V_{vw} and reverse the sign of both dV/dh and V in the direction of repulsion. On further reduction of the gap (very small h), V_{vw} should again predominate, since $V_{el} \to 64cRT\gamma^2/\varkappa$ at $h \to 0$, whereas V_{vw} increases indefinitely when $h \to 0$*. There is thus a repulsion maximum in the function $V = V(h)$, which can be easily found from the condition $\partial V/\partial h = 0$. The choice of solution (maximum or minimum) does not present any difficulty since V is positive for the maximum.

The solution of (7.30) with respect to V_{max} is cumbersome. It is, however, readily appreciated that when the electrolyte concentration is increased, the magnitude of $\varkappa$ in the exponent of V_{el} is also increased (compression of diffuse layers), so that the maximum caused by it becomes lower. At a certain value of c the curve $V(h)$ will become similar to curve b in Fig. 61 with $V_{max} = 0$. In accordance with all that has been said before, coagulation will become fast starting from this concentration. This is therefore the critical concentration, c_{cr}. In other words, the critical concentration can be calculated from simultaneous solution of

$$\partial V/\partial h = 0 \quad \text{and} \quad V = 0. \qquad (7.31)$$

When written out in full, indexing all magnitudes related to h and c with subscript "cr" (critical), this becomes

* This statement is still of formal nature since the relationship $\Pi_{vw} = -K/h^3$ becomes unreliable for very thin layers, as already indicated during its derivation; for monolayers the adsorption is always positive, so that on transition to a monolayer Π_{vw} should undergo a reversal. Extrapolation for Π_{el} is also not permissible in the case of very thin layers, where the actual conditions do not satisfy the prerequisites of theory. For these reasons the boundary transitions used are arbitrary and are applicable in cases when V_{max} arises at a sufficient distance between particles.

$$\left.\begin{aligned} \frac{\partial V}{\partial h} &= -64c_{cr}RT\gamma^2\, e^{-\varkappa_{cr}h_{cr}} + \frac{K}{h_{cr}^3} = 0 \\ \text{and} \qquad V &= \frac{64c_{cr}RT\gamma^2}{\varkappa_{cr}}\, e^{-\varkappa_{cr}h_{cr}} - \frac{K}{2h_{cr}^2} = 0 \end{aligned}\right\} \tag{7.32}$$

Dividing these equations by each other we obtain $h_{cr} = 2/\varkappa_{cr}$, which, after substitution into one of them and replacing $\varkappa_{cr}$ by $\sqrt{8\pi z^2F^2c_{cr}/\varepsilon RT}$, yields

$$c_{cr}z^6 = \text{const.}, \tag{7.33}$$

where the constant includes all magnitudes except z and c_{cr}^*.

Expression (7.33) means that the critical concentrations of ions of various valencies are inversely proportional to the sixth power of their valency z, *i.e.*

$$c_{cr1}:c_{cr2}:c_{cr3}:c_{cr4} = 1:(\tfrac{1}{2})^6:(\tfrac{1}{3})^6:(\tfrac{1}{4})^6 = 1:0.016:0.0014:0.00024,$$

which corresponds fairly accurately to the Schulze–Hardy rule.

Tables VII, 4 and 5 contain the results collected by Overbeek[23] from investigations of various authors** concerning the measurements of c_{cr} for the coagulation of negatively and positively charged sols. Although the individual values are scattered in view of the low accuracy of the methods available for the determination of c_{cr}, the mean values are in good agreement with (7.33).

Until recent times this has been the fundamental proof of the theory of coagulation of lyophobic sols, referred to before, and developed by Deryagin, Verwey and Overbeek[24].

More detailed checking and utilisation of this theory is made difficult by the complex nature of the coagulation process and by the mathematical difficulties encountered when theory is applied to the more real and less simplified cases (*e.g.* to spherical particles of unequal size).

Some new aspects and possibilities of this theory have emerged from the work of Glasmann, Dikmann *et al.* (1955)[29]. These authors investigated the coagulating effect of a mixture of electrolytes, and demonstrated theoretically that in the case of two symmetrical electrolytes in which the valencies of the gegenions differed by a factor of 2, a clear synergic effect should be observed. Analogous calculations of Π_{el} for other types

* The dependence of γ on z is neglected here. This is permissible when the φ_0 potential is sufficiently high.

** Sols of As_2S_3, Au and Fe_2O_3 — Freundlich (1903–1910)[26], the AgI sol — Kruyt and Klompé (1943)[27], and the Al_2O_3 sol — Ishizaka (1913)[28].

References p. 240

TABLE VII, 4

CRITICAL CONCENTRATIONS FOR THE NEGATIVELY CHARGED SOLS
(in m moles/l)

Electrolyte	As_2S_3 *sol*	*Mean value*	Au *sol*	*Mean value*	AgI *sol*	*Mean value*
LiCl	58		—		—	
$LiNO_3$	—		—		165	
NaCl	51		24		—	
$NaNO_3$	—		—		140	
KCl	49.5		—		—	
KNO_3	50	55	25	24	136	142
$^1/_2$ K_2SO_4	65.5		23		—	
$RbNO_3$	—		—		126	
Potassium acetate	110		—		—	
HCl	31		5.5		—	
$AgNO_3$	—		—		0.01	
Morphine hydrochloride	0.42		0.54		—	
Fuchsine NB	0.11		0.002		—	
$MgCl_2$	0.72		—		—	
$Mg(NO_3)_2$	—		—		2.60	
$MgSO_4$	0.81		—		—	
$CaCl_2$	0.65		0.41		—	
$Ca(NO_3)_2$	—		—		2.40	
$SrCl_2$	0.635	0.69	—	0.38	—	2.43
$Sr(NO_3)_2$	—		—		2.38	
$BaCl_2$	0.69		0.35		—	
$Ba(NO_3)_2$	—		—		2.26	
$ZnCl_2$	0.685		—		—	
$Zn(NO_3)_2$	—		—		2.50	
$UO_2(NO_3)_2$	0.64		2.8		3.15	
$CuSO_4$	—		0.015		—	
$Pb(NO_3)_2$	—		0.002		2.43	
Quinine sulphate	0.24		—		—	
Benzidine nitrate	0.09		—		—	
$AlCl_3$	0.093		—		—	
$Al(NO_3)_3$	0.095		—		0.067	
$^1/_2Al_2(SO_4)_3$	0.096	0.091	0.009	0.006	—	0.068
$La(NO_3)_3$	—		—		0.069	
$Ce(NO_3)_3$	0.080		0.003		0.069	
$Th(NO_3)_4$	0.090	0.090	0.0009	0.0009	0.013	0.013

TABLE VII, 4 *(continued)*

MEAN CRITICAL CONCENTRATIONS FOR THE COAGULATION OF NEGATIVELY CHARGED SOLS

Valency of the coagulating ion	*Mean critical concentration for the* As_2S_3 *sol*	*Ratio*	*Mean critical concentration for the* Au *sol*	*Ratio*	*Mean critical concentration for the* AgI *sol*	*Ratio*	*Theoretical ratio*
1	55	1	24	1	142	1	1
2	0.69	0.013	0.38	0.016	2.43	0.017	0.016
3	0.091	0.0017	0.006	0.0003	0.068	0.0005	0.0013
4	0.090	0.0017	0.0009	0.00004	0.013	0.0001	0.00024

TABLE VII, 5

CRITICAL CONCENTRATIONS FOR POSITIVELY CHARGED SOLS

(in m moles/l)

Electrolyte	Fe_2O_3 *sol*	*Mean value*	Al_2O_3 *sol*	*Mean value*
NaCl	9.25	11.8	43.5	52
KCl	9.0		46	
NH_4Cl	—		43.5	
$^1/_2BaCl_2$	9.65		—	
KBr	12.5		—	
KI	16		—	
KCNS	—		67	
KNO_3	12		60	
$^1/_2Ba(NO_3)_2$	14		—	
HCl	> 400		—	
$^1/_2Ba(OH)_2$	0.42		—	
Sodium picrate	—	0.21	4.7	0.63
K_2SO_4	0.205		0.30	
Tl_2SO_4	0.22		—	
$MgSO_4$	0.22		—	
$MgCrO_4$	—		0.95	
$K_2Cr_2O_7$	0.195		0.63	
Potassium oxalate	—		0.69	
H_2SO_4	~ 0.5		—	
$K_3Fe(CN)_6$	—		0.080	0.080
$K_4Fe(CN)_6$	—		0.053	0.053

References p. 240

TABLE VII, 5 *(continued)*

MEAN CRITICAL CONCENTRATIONS FOR POSITIVELY CHARGED SOLS

Valency of the coagulating ion	Fe_2O_3 *sol*	*Ratio*	Al_2O_3 *sol*	*Ratio*	*Theoretical ratio*
1	11.8	1.00	52	1.00	1.00
2	0.21	0.018	0.63	0.012	0.016
3	—		0 .080	0.0015	0.0013
4	—		0.053	0.0010	0.00024

of electrolyte mixtures resulted in the theoretical prediction of the opposite effect—antagonism* [Glasmann, Dikmann, Streltsova (1957)[30]]. The experimental checks carried out by the coagulation of rosin hydrosols confirmed these calculations [Glasmann, Zhelvis (1962)[31]]. In 1961 Barboi, Glasmann and Dikmann[32] also put forward a more complex theory of the effect of electrolyte mixtures, which allowed for the adsorption phenomenon and made it possible to determine the critical concentration of the mixture from the critical concentrations and adsorbabilities of the components.

In view of the difficulties encountered in the direct checking of the theory, the present author believes in the importance of indirect checks with the aid of various simplified models. The basic feature of the theory is the commensurability of the effects of the electrostatic and van der Waals components of the interaction energy (or interaction forces). In this respect the measurements carried out by Sheludko and Ekserova (1960)[33] with free thin layers of aqueous electrolytes, described in the preceding chapter, provide considerable support to the theory of Deryagin, Verwey and Overbeek. An even closer approach to the conditions of coagulation could be provided by measuring interactions arising between condensed phases. Such model investigations have not as yet yielded results which could be unambiguously and clearly compared with the theory. It can, however, be stated with certainty that, after overcoming

* When a mixture of electrolytes is used in coagulation, their effects can be additive, super-additive (synergism) (x% of one coagulating concentration and y% of another, $x + y > 100$) or antagonistic ($x > 100$ or $y > 100$).

the considerable experimental difficulties, these studies will yield highly important results.

5. Model Studies of Coagulation

The first experiments in this field were carried out by Buzágh who, starting from 1929[34], investigated the adhesion of suspended particles to a smooth surface. A suspension of quartz in a closed cell with a plane-parallel quartz bottom was allowed to settle for some time, and the number of particles which had sunk to the bottom was counted, using a microscope. The cell was then carefully inverted and the number of particles which did not fall off was determined by the same method. The relative number of particles which adhered to the bottom (the adhesion number) is a quantitative measure of the process of adhesion. The disadvantage of this method consists in the fact that the force which causes the detachment is determined by the weight of particles in the solution, whereas the force of adhesion is proportional to the area of the surface of contact. With increasing particle size both these opposing factors increase, although not to the same extent. The adhesion number is thus observed to pass through a maximum when the particle size is increased, as may be seen from Table VII, 6.

In spite of this specific feature, the method could be used to demonstrate clearly the effect of the ζ-potential on adhesion (Table VII, 7). By using the same method it was possible to demonstrate that the existence of two zones of stability is associated with a reversal of the sign of

TABLE VII, 6

ADHESION NUMBERS FOR SUSPENSIONS OF QUARTZ PARTICLES

Particle diameter (in μ)	*Adhesion number*
1	8.5
1–2	78.2
2–3	97.7
7–8	22.3
12	9.4
20–21	6.5
32	0

References p. 240

TABLE VII, 7

THE RELATIONSHIP BETWEEN ADHESION NUMBERS AND THE ELECTROPHORETIC MOBILITY OF PARTICLES

Salt concentration (m equiv./l)	*Adhesion number*			*Electrophoretic velocity* (μ/sec)		
	$CaCl_2$	$SrCl_2$	$BaCl_2$	$CaCl_2$	$SrCl_2$	$BaCl_2$
0	46.2	46.2	46.2	2.09	2.09	2.09
0.0125	65.8	73.2	86.7	—	—	—
0.025	89.8	92.5	100.0	—	—	—
0.05	100.0	100.0	100.0	1.80	1.80	1.45
0.1	100.0	100.0	100.0	1.72	1.72	1.38

the sol charge. Thus *e.g.* in the case of ions Al^{3+} and Th^{4+} which reverse the charge on quartz, the adhesion numbers increased with increasing concentration up to the disappearance of ζ, after which the adhesion numbers decreased again as the ζ-potential increased when the charge was reversed. In another, rather more detailed method, Buzágh also determined the degree of adhesion. When the sedimentation is complete, the cell, which has a quartz bottom and which is filled with a suspension of quartz, is gradually tilted and the settled particles are observed under a microscope. When the cell is inclined the particles are acted upon by forces of "shear", which increase with the angle of inclination, and cause the particles to slide. Depending on the degree of adhesion, the sliding begins at a certain angle of inclination, called by Buzágh the "angle of sliding".

The model studies of Deryagin and Obukhov (1935)[35] and of Deryagin and Kusakov (1937)[36] are considerably more rigorous. The method used by these authors consists in pressing two plates (immersed in a liquid) against each other with a certain force. The liquid is then sucked into the gap and draws the plates apart, until the (positive) disjoining pressure in the layer is equal to the external pressure. An equilibrium layer will then form, whose thickness is determined mechanically by the displacement of the plates. Measurements were carried out with mica plates in distilled and twice-distilled water and in electrolyte (NaOH) solutions, and with steel plates in paraffin oil.

These noteworthy experiments, which initiated the studies of the disjoining pressure and in general of the specific thermodynamic properties of thin layers, unfortunately could not be interpreted theoretically, because no theory of these phenomena existed at that time. Even today their quantitative interpretation on the basis of the Deryagin–Landau theory of Π_{el} remains extremely difficult, owing to the imprecise knowledge of electrolyte concentration in the case of low concentrations where this theory can be applied with success. Approximate calculations show that the measured maximum equilibrium thickness of the order of 1–2 μ (at a pressure of 4.4 g/cm^2) are improbably large. It is, however, highly interesting, and wholly natural, that the disjoining pressure was not detected (the equilibrium thickness was zero) in 0.03% NaOH. At high solution concentrations, considerable equilibrium thicknesses were found both for water and for paraffin oil, which is a mixture of hydrocarbons. This observation cannot be explained by the elementary theory, but may be related to the positive disjoining pressure in thin layers of concentrated solutions which still awaits a quantitative explanation.

After a protracted period of inactivity, model studies of this type were resumed in 1959–1962 by Voropaeva, Deryagin and Kabanov[37], who used a new highly ingenious method of crossed filaments. In essence, the experiment consisted in very slowly bringing a platinum wire nearer and nearer to another platinum wire immersed in a liquid, until they touched. In the absence of any interaction between the wires, the fixed filament would not be displaced up to the moment of contact (which can be very accurately determined electrically). In actual fact, however, a jerk is frequently observed prior to the contact, indicating the presence of repulsion forces (equivalent to a positive Π). The "force barrier" of repulsion can be evaluated from the magnitude of maximum deviation before the contact. When the wires are polarised with respect to a certain reference electrode and reduced to the zero-charge point, then, as expected, a slight repulsion corresponding to the removal at that instant of the diffuse electric layers and of Π_{el} can be observed in aqueous electrolytes. The zero-charge point may thus be established by this method.* It is interesting to note that in many cases a certain residual repulsion was observed at that point, and again this effect cannot be explained theoretically. The crossed filaments method also allowed the detection of $+\Pi$ in some concentrated

* The interpretation of the potential corresponding to the minimum repulsion as the zero-charge point is not obvious.

solutions, in which, at roughly the same concentrations, Sheludko and Ekserova (1960)[38] observed positive disjoining pressure in microscopic free films.

Direct measurements of the interaction forces between solids in a liquid were also attempted by Popovski (1962)[39].

On the basis of the above results, we may expect that perfected model studies of this kind will ultimately not only make it possible to check the Deryagin–Verwey–Overbeek theory in a quantitative manner and to determine the important constants K and φ_0, but will also provide an experimental basis for the further development of this theory.

It should be noted that, on the basis of experimental observations of cases in which the detected positive disjoining pressure cannot be explained from the standpoint of the present theory, Deryagin introduced the concept of "forces of the third kind", related to an assumed specially organised polymolecular solvation layer of the dispersion medium at the interface. It is evident that neither the molecular nor the microscopic theory of the effects of van der Waals interaction can help in elucidating this concept since both these theories treat thin layers as structureless formations. The idea of the structure of the solvation layer and the ideas of the special behaviour of thin layers of solutions, in which the concentration is affected by long-range van der Waals forces present, are in the author's opinion, the newest and most disputed trends in the field under consideration. The ingenious idea of Martynov[40], according to which the energy in a thin layer of a solution can change owing to a change in its fluctuation structure, which is deformed in a sufficiently thin layer, is also noteworthy. This is really an improvement on the explanation proposed for the phenomena, by Sheludko and Ekserova (1959)[41], according to which the "diffusely" organised solvation of molecules in solutions is altered in a thin layer.

REFERENCES

1 Hardy, *Proc. Roy. Soc. (London)*, 66 (1900) 110; *Z. Phys. Chem.*, 33 (1900) 385.
2 Powis, *Z. Phys. Chem.*, 89 (1914) 145.
3 Schulze, *J. Prakt. Chem.*, 25 (1882) 431; 27 (1883) 320.
4 Freundlich, *Kapillarchemie; eine Darstellung der Chemie der Kolloide und verwandte Gebiete*, Akad. Verlagsges., Leipzig, 1932, Bd 2, p. 122.
5 Lepin and Bromberg, *Zhur. Fiz. Khim.*, 13 (1939) 71.
6 Freundlich, *Kolloidchem. Beihefte*, 22 (1926) 95; *Z. Phys. Chem.*, 129 (1927) 305.

7 KARGIN, *Acta Physicochim. URSS*, 3 (1935) 97; 8 (1938) 675.
8 SMOLUCHOWSKI, *Z. Phys. Chem.*, 92 (1918) 129.
9 MÜLLER, *Kolloid-Z.*, 38 (1926) 1.
10 WIEGNER AND TUORILA, *Kolloid-Z.*, 38 (1926) 3.
11 MÜLLER, *Kolloidchem. Beihefte*, 26 (1928) 257.
12 WIEGNER AND MARSHALL, *Z. Phys. Chem.*, A 140 (1929) 1, 39.
13 SMOLUCHOWSKI, *Z. Phys. Chem.*, 92 (1917) 155.
14 TUORILA, *Kolloidchem. Beihefte*, 24 (1927) 1.
15 MÜLLER, *Kolloidchem. Beihefte*, 27 (1928) 223; LEVICH, *Dokl. Akad. Nauk, SSSR*, 99 (1954) 809.
16 ZSIGMONDY, *Z. Phys. Chem.*, 92 (1917) 600.
17 WESTGREN AND REITSTÖTTER, *Z. Phys. Chem.*, 92 (1917) 750.
18 KRUYT AND VAN ARKEL, *Rec. Trav. Chim.*, 39 (1920) 656; 40 (1921) 169.
19 TUORILA, *Kolloidchem. Beihefte*, 22 (1926) 191.
20 FUCHS, *Z. Phys. Chem.*, A 171 (1934) 199.
21 LANGMUIR, *J. Chem. Phys.*, 6 (1938) 873.
22 HAMAKER, *Physica*, 4 (1937) 1058.
23 KRUYT (Editor), *Colloid Science*, Elsevier, Amsterdam, 1952, Vol. 1.
24 VERWEY AND OVERBEEK, *Theory of the Stability of Lyophobic Colloids*, Elsevier, Amsterdam, 1948; DERYAGIN, *Discussions Faraday Soc.*, 18 (1951) 85.
25 FRENKEL', *Kinetic Theory of Liquids*, Dover, New York, 1955.
26 FREUNDLICH, *Z. Phys. Chem.*, 44 (1903) 129; 73 (1910) 385.
27 KRUYT AND KLOMPÉ, *Kolloidchem. Beihefte*, 54 (1943) 484.
28 ISHIZAKA, *Z. Phys. Chem.*, 83 (1913) 97.
29 GLAZMAN AND DYKMAN, *Dokl. Akad. Nauk, SSSR*, 100 (1955) 299; *Kolloidn. Zh.*, 18 (1956) 13.
30 GLAZMAN, DYKMAN AND STREL'TSOVA, *Dokl. Akad. Nauk, SSSR*, 117(1957) 829; *Kolloidn. Zh.*, 20 (1958) 149.
31 GLAZMAN AND ZHEL'VIS, *Kolloidn. Zh.*, 24 (1962) 405.
32 BARBOI, GLAZMAN AND DYKMAN, *Kolloidn. Zh.*, 23 (1961) 376, 381.
33 SHELUDKO AND EKSEROVA, *Kolloid-Z.*, 165 (1959) 148; 168 (1960) 24.
34 BUZAGH, VON, *Kolloid-Z.*, 47 (1929) 370; 51 (1930) 105, 230.
35 DERYAGIN AND OBUKHOV, *Kolloidn. Zh.*, 1 (1935) 385.
36 DERYAGIN AND KUSAKOV, *Bull. Acad. Sci. URSS, Classe Sci., Math., Nat., Sér. Chim.*, (1937) 1119; *Acta Physicochim., URSS*, 10 (1939) 25, 153.
37 VOROPAEVA, DERYAGIN AND KABANOV, *Kolloidn. Zh.*, 24 (1962) 396.
38 SHELUDKO AND EKSEROVA, *Izv. Khim. Inst. Bulgar. Acad. Nauk*, 7 (1960) 105.
39 POPOVSKII, *Pyataya Vsesojuziaya Konferenciya po Kolloidnoj Khimii* (*Fifth All-Union Conference on Colloid Chemistry, Theorems*), Odessa, 1962, p. 25.
40 MARTYNOV, Personnal communications.
41 SHELUDKO AND EKSEROVA, *Annuaire Univ. Sofia*, 54 (1959/60) 205.

CHAPTER VIII

Foams and Emulsions

Disperse systems consisting of a gas in a liquid or of a liquid in another liquid are known respectively as foams and emulsions. The distinguishing feature of these systems, as opposed to the dispersions of solid particles in a liquid (sols) is the liquid consistency of interfaces. The particle shape is therefore determined by the condition of minimum surface for a given volume. In dilute foams and emulsions the particles are spherical, while in concentrated ones the particles are deformed by mutual pressure, forming polyhedral shapes of a definite type (dodecahedra in the case of monodisperse systems). The process of the destruction of the disperse system does not in this case end in the agglomeration of particles (coagulation), but can go further until complete coalescence of the particles is attained.

Dilute foams and emulsions are similar to sols. The problem of stability is determined, just as in the case of sols, by the phenomena of sedimentation or creaming and coagulation, and if we again confine ourselves to analysing the stability of dilute systems, it is immaterial whether the "precipitate" is disperse (though coagulated) or whether it has undergone coalescence.

A discussion will now be given of some specific characteristics of foams and emulsions, *e.g.* the fact that since the two phases are similar in the case of emulsions, each liquid can be dispersed in the other.

The fundamental difference between sols and concentrated foams and emulsions lies in the possibility of coalescence in the latter case. The problem of the stability of these systems is rather specific and deserves detailed consideration. It is obvious that it is now unnecessary to study the problem of the collision frequency, which is of great importance in dilute disperse systems, since in foams and emulsions the particles are tightly packed and are in continuous contact. For the same reasons, it is immaterial in this case whether attraction forces which bring about agglomeration exist between the particles. On the other hand, we now encounter the highly important problem of the resistance of the thin layers, separating the gaseous or liquid particles, to rupture which brings about coalescence. The hydrodynamic factor, which was manifested by sedimentation and diffusion of particles in the dilute disperse systems, becomes here the problem of the drainage of liquid from thin layers as a result of hydrostatic pressure, capillary forces, and the disjoining pressure. The problem of the stability of concentrated foams and emulsions is thus reduced to the investigation of (*a*) the nature and cause of the thinning of liquid walls of these cell-like structures, and (*b*) the nature and cause of their sudden rupture (together with the layer thicknesses involved). Such a detailed approach to the stability of concentrated foams and emulsions has not, unfortunately, been consistently applied. Various theories aiming at reducing the stability to one single factor or mechanism were instead suggested, which have so far failed to elucidate the problem. The science of foams and emulsions, in which the stability problem occupies the predominant position, is thus characterised by rich and diverse experimental material, by the complete absence of a consistent theory, and by the extremely contradictory nature of the individual attempts at a theoretical explanation of the experimental facts. Some idea about this state of affairs can be obtained from the detailed accounts of Clayton[1] and Bikerman[2].

In recent times results of investigations aiming at a generalization of the available data have been published in the field of foams[3,4,5], which apparently lends itself more readily to model studies. This generalisation can now be supplemented by new results[6]. It can also be hoped that this approach will lead not only to a more complete understanding of foams, but also to the establishment of a theory of the stability of emulsions. The resemblance of these two systems should make it possible to extend the theories developed for foams, and the experimental methods of investigation, also to the case of emulsions.

For these reasons we shall confine ourselves in this chapter to the minimum of experimental material collected earlier for these systems, concentrating almost completely on the data, observations and conclusions, which, on the basis of model experiments, have prompted an analysis of the various factors and processes occurring in foams. The final conclusions which can now be reached in this way do not, however,

References p. 273–274

exhaust the problem under investigation. Special literature mentioned previously contains a wealth of additional information, which may be utilised by readers who wish to take this generalisation further, or to apply the facts to practical cases. The more general considerations introduced in the present book should, however, familiarise the reader with, and permit him to appreciate the actual facts; this is the idea on which the present rather unusual method of presenting the study of foams and emulsions has been based.

1. The Structure of Foams

The present section will be confined to the case of concentrated foams (which will from now on be referred to merely as foams), and will contain a brief review of the phenomenological material relating to these systems.

Foams may be formed by shaking a liquid or by passing through it a current of gas, and appear as a column or a layer on the surface of the liquid. Foams may have various structures, and unless special precautions are taken, they are polydisperse. The individual bubbles pressed against each other are separated by very thin, approximately plane-parallel films, and by thicker regions of liquid in places where an individual film is in contact with another film, with the bulk of the liquid, or with the wall of the vessel. In view of the concave shapes of these thicker regions, a capillary pressure exists between them and the plane portions (films), which tends to drain the liquid from the films into the thicker portions. Moreover, since in polydisperse foams the curvature of the thickened "ribs" varies, capillary transfer of liquid may take place from some thicker parts (with smaller curvature) to others. These thickened and concave formations which drain the liquid from the films are often known as "Plateau borders" or "Gibbs' triangles". In a column of foam we also encounter hydrostatic pressure, which causes the liquid from the foam to flow into the underlying solution. As a result of these two factors, foams gradually lose some of the liquid contained in them (a syneresis characteristic of foams), which leads to a thinning of the films and to an increase in the curvature of the concave sections. When the films become sufficiently thin ($h \approx 10^{-5}$ cm), an additional factor comes into play, *i.e.* the disjoining pressure, which is usually negative* (*cf.* chapter VI) and which promotes further reduction of the film thickness.

The bubbles on the upper surface of a column of foam are in a special state, since they are not fed by the liquid flowing down from the upper layers; furthermore, the draining hydrostatic pressure has here a maximum value. The films are therefore thinnest in these upper layers, and the process of breakdown occurs in unstable foams mainly from above.

The instability of a foam column is determined not only by syneresis but also, although to a lesser extent, by the fact that the gas in bubbles is compressed by capillary forces and may diffuse across the films into the free gaseous phase; the mechanism of the latter process was analysed by de Vries[3].

It is therefore clear that the foam loses both the liquid and the gaseous phases as a result of the transfer phenomena. However, these phenomena are not a direct cause of the breakdown of foam, which is achieved by film rupture, but only bring the foam to this final state.

The possibility of foam formation is thus determined by the possibility of the formation of foam films. It was, however, already shown in Chapter VI that a foam film may form as a certain unstable structure only when the liquid contains a surface-active component, which retards the flow in the film surface and the corresponding hydrodynamic drainage. Pure liquids and solutions which contain no surface-active components cannot form films of any stability and consequently cannot be foamed. Although certain authors believe that weak foams may be obtained even in the absence of surface-active agents, these opinions should be considered as erroneous. Such conclusions may be due to the fact that foam-formation is promoted by extremely small quantities of surface-active substances, which might, unknown to the experimenters, have been present as an impurity.

2. Role of Surface-active Agents

The surface-active component is therefore the most important factor in the formation of foam. Not all surface-active agents promote foam formation to the same extent, and their effectiveness is assessed by means of the so-called foam stability. This rather vague concept becomes clearer if we introduce a certain experimental method of measuring this property

* When the electrolyte concentration is sufficiently high to repress Π_{el}.

References p. 273–274

of foam. The methods suggested so far, are unfortunately by no means perfect, and consist in the measurement of foam persistence, τ.

In the frequently used method systematically applied by Bartsch (1924)[7], a small volume of the liquid under examination is shaken in a closed test-tube, keeping the conditions as standard as possible, until the maximum amount of foam has been formed. The persistence of the resultant foam column is then measured. It is recommended that the time measured should be taken to the point at which a "mirror", *i.e.* free surface, can be observed. This procedure does not take into account the longest lifetimes of the last foam bubbles (especially in the case of the most resistant foams), which surround the free surface, but reduces the time of measurements and improves their reproducibility, which is fairly good when the mean of 10 to 15 individual measurements is taken. On the other hand, this practice tends to mask the effect of the stabiliser (particularly in the case of medium stabilities), which is most readily observed on these last bubbles. Similar remarks can be applied to the measurements of foam half-life, *i.e.* of the time during which the foam column is reduced by one-half. This method is characterised by unsatisfactory reproducibility in the case of very persistent foams.

Another method, suggested by Hardy in 1927[8], consists in determining the lifetime of a bubble formed on a free surface. This model method yields well-reproducible results, and may also be applied to the determination of the persistence of emulsions, in which case the lifetime of an individual droplet is measured. It is unfortunately still uncertain to what extent the lifetimes of the bubble and of the foam column coincide with one another. The point is that the stability of a foam column depends to a certain extent on its initial structure, and on the change of this structure when the foam is breaking down. The foam structure is not, however, the same for the various frothing agents and for various concentrations of such agents. No systematic investigations of this problem are at present known to the author. Therefore it is still extremely difficult to decide just how reliably the persistence of an individual bubble or of a microscopic or macroscopic free film may be taken as a measure of foam stability.

Certain other, more complicated methods have been put forward, such as the stationary column method. In this procedure, air is bubbled through the liquid at a constant rate, until a foam of constant height is formed for which the rates of formation and destruction are exactly balanced. This method is suitable for comparing foams of similar stability, but cannot be used for investigating foams of widely differing stabilities, since this would necessitate a change in the design of the apparatus. A further disadvantage lies in the fact that the results obtained cannot be easily compared with the more conventional foam persistence. It is possible that if this method is standardised to a certain extent, it will be possible to find for it a wider application in the future.

Systematic investigations of the effects of the nature and concentration of the foaming agent on the stability of the resultant foams were carried out by Bartsch (1924)[7] by the first-mentioned method. It was found that foams can be resolved into two, widely differing groups: the unstable foams, stabilised by low-molecular frothers (low fatty alcohols and acids) with persistencies ranging from a few seconds to 20 seconds, and the stable foams, stabilised by detergents (surface-active soap-like substances), with

TABLE VIII, 1

CHARACTERISTIC VALUES OF UNSTABLE FOAMS

Aqueous solutions	*Maximum foam persistence* τ (sec)	*Optimum concentration* (moles/l)
Alcohols		
ethyl	5.0	0.28
propyl	11.0	0.34
isobutyl	12.0	0.09
isoamyl	17.0	0.036
tert.-amyl	10.0	0.034
heptyl	8.0	0.00007
octyl	5.0	0.0003
Acids		
formic	4.0	0.45
acetic	8.0	0.20
propionic	11.0	0.25
butyric	18.0	1.00
valeric	9.0	0.015
caproic	13.0	0.0075
heptanoic	16.0	0.0015
caprylic	12.0	0.00025
pelargonic	5.0	0.00007

References p. 273–274

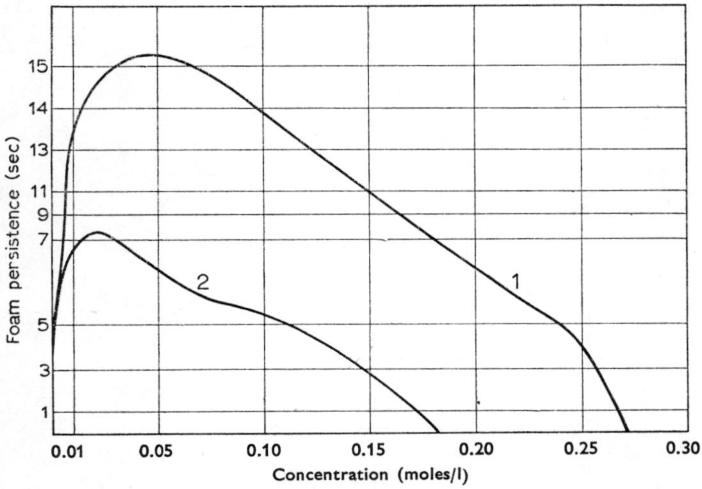

Fig. 62. Dependence of foam persistence on the concentration of the foaming agent: (1) aqueous solution of isoamyl alcohol; (2) aqueous solution of *m*-cresol.

persistencies measured in hours or even days. The unstable foams are characterised by the fact that the variation in their duration, τ, with concentration c of the surface-active agent has a clearly expressed maximum (Fig. 62).

The values of maximum τ and the corresponding optimum concentrations are listed in Table VIII, 1 for a number of low-molecular surface-active substances. The effect of the concentration of electrolytes on τ was found to be rather small. At any rate, one cannot observe in foams anything similar to the strong effect which electrolyte concentration exerts on the stability of lyophobic sols. Bartsch's measurements showed further that the viscosity of the solution (increased by adding sugar) had only a limited effect on τ: thus for a syrup (50 % of sugar) τ increased by only 10 %.

A transition from small to large τ can be observed when working with solutions of detergents in cases when the detergent is sufficiently stable in very low concentrations corresponding to τ-values expressed in seconds. Such investigations were carried out by Ekserova[6] (1962) with a large number of foaming agents, and showed that the function $\tau(c)$ always possesses a clearly expressed S-like form (Fig. 63). The gradual increase of τ, with the rising concentration c of the detergent, eventually becomes

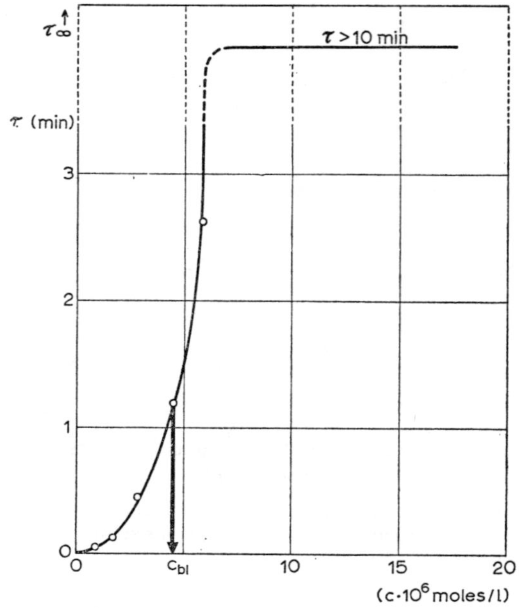

Fig. 63. Dependence of foam persistence τ on the concentration of OP20 (octylphenol condensed with polyoxyethylene, and containing 20 monomer units; aqueous solution +0.1 mole KCl per litre). c_{bl} is the concentration at which black spots begin to form.

very steep, indicating the presence of a new, special mechanism of the stability characteristic of longer foam lifetimes.

In 1927[8] Hardy studied the persistence of single bubbles during their stabilisation by means of an insoluble stretched-type monolayer. In this case, very small values of τ are always obtained, and therefore these investigations relate mostly to the stability problem of the short-lived foams. Just as in the case of low-molecular, soluble, surface-active stabilising agents, it was found that τ passed through a maximum corresponding to surface concentrations similar to, but lower than, the concentration of the saturated monolayer.

Analogous investigations into the monolayers of the normal (non-stretched) type carried out by Talmud and Sukhovol'skaya in 1931[9], similarly revealed the existence of a τ maximum at surface concentrations lower than those corresponding to close packing. Analogous results were also obtained later by Trapeznikov[10]. The validity of measurements carried out by Talmud and Sukhovol'skaya has been doubted by Adam[11], who argued that non-stretched monolayers at concentrations below (but close to) the saturation concentration are a two-dimensional heterogeneous formation with a horizontal section in the $\sigma(c)$ curve, and consequently with $d\sigma/dc = 0$. According to Adam, maximum stabilisation should occur at maximum $d\sigma/dc$. The present author believes, however, that it would be more correct to draw the opposite conclusion, *i.e.* that Adam's assumption about the role of $d\sigma/dc$ is wrong. A more detailed account of Gibbs' theory, on which Adam based his conclusion, will be given later.

Mention should also be made of the interesting observations and investigations carried out by Perrin and his co-workers[12] on "black films". If a separate free film or bubble formed from a sufficiently concentrated solution of Castile soap (whose main component is sodium oleate) is observed, it is readily noticed that after its formation the film gradually becomes thinner, changing its interference colour after reaching a thickness of ~ 1000 Å, when the interference colour passes into ultraviolet and the film appears to be white. On further decrease of thickness the amount of reflected light is decreased, the white colour appears darker, and the film becomes grey and of non-uniform thickness. It is of interest that this non-uniformity proceeds in steps for the concentrated soap solutions: individual portions appear in different shades of grey, attaining full black colour in the thinnest portions. In 1921 Wells[13] carried out a systematic series of measurements of the thickness of such "stratified" films, by means of interferometric and colorimetric methods, and found that in agreement with Perrin's assumption, the thicknesses of black films were multiples of the thickness of the thinnest (and therefore the blackest) portions of the film. The minimum thickness was 42–45 Å, corresponding approximately to twice the length of the sodium oleate molecule.

In recent times Sheludko and Ekserova[6] (1961) on the one hand, and Duyvis (1962)[14], on the other, carried out corresponding accurate measurements of the thickness of the thinnest black films obtained with several detergents, confirming the earlier results. It was found that the thinnest films from sodium oleate solutions have a thickness of 40 Å, whilst values of 60–70 Å and 85–100 Å, respectively, were found for OP7 and OP20, these values being also of the order of twice the molecular length of these detergents (36 Å and 67 Å). Such small thicknesses were, however, only obtained in the presence of sufficient amounts of electrolyte in the solution. When the electrolyte content was too low, thicker films

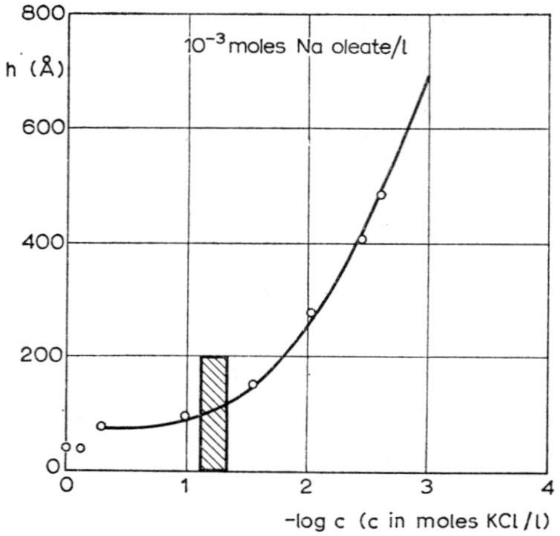

Fig. 64. Dependence of the final (equilibrium) thickness h_r of microscopic films from aqueous solutions of sodium oleate (10^{-3} mole/l) on the concentration of KCl. Actual measurements are shown by small circles; the curve is calculated from theory. The hatched region corresponds to concentrations at which the thinning before the equilibrium thickness has been reached ceases to be smooth, and takes place with the formation of black spots. The first two points correspond to a Perrin film with a thickness of 40 Å.

formed on account of Π_{el}, whose equilibrium thickness decreased smoothly with increasing electrolyte content, in agreement with the Deryagin–Landau theory of Π_{el}*. The transistion from the film stabilised by Π_{el} to the doubled adsorption layer—Perrin's lamella of first order—was manifested in the case of sodium oleate by a sudden discontinuity in the film thickness/electrolyte concentration curve at the KCl concentration of 0.5 to 0.8 moles/l (Fig. 64). These investigations have also explained the apparent contradiction with the data obtained by Deryagin and Titievskaya (1953)[15] for the thickness of black films. The latter authors used electrolyte concentrations which were insufficient to suppress Π_{el} and black films with a greater equilibrium thickness were formed in place of Perrin's lamellae.

The following note may perhaps be added with regard to black films. Doubled adsorption layers or the multimolecular films in equilibrium due to Π_{el} appear black, irrespective of their structure, when their thickness is less than 300–200 Å.

* With due allowance for Π_{vw}.

In certain cases (particularly in the case of sodium oleate) the black Perrin films are particularly strong and persistent. Thus Dewar (1917–1923)[16] succeeded in preserving a black film for over 3 years in a closed vessel. The mechanism of the rupture of such doubled adsorption layers has recently (1962) been the subject of a theoretical investigation by Deryagin and Gutop[17]. Assuming these formations to behave like two-dimensional crystals, the latter authors succeeded in developing quantitatively a "cavitation" or "hole" mechanism of film breaking, following the ideas of de Vries[3] (*cf.* Chapter VI). Since in this case the thin layer cannot break down by the deflection mechanism (which is more convenient from the point of view of energy in a multimolecular layer) the hole mechanism can be applied; this is in agreement with the high durability associated with the very small thickness of Perrin films. It has been shown before, however, that black Perrin films are not the only ones, nor even the most frequently occurring types of black films, so that the problem of the stability of Perrin lamellae by no means exhausts the problem of the stability of black films and foams.

3. Theories of Foam Stability

There are at present three different points of view on the basis of which a general explanation of foam stability can be attempted, none of which can be regarded as either proved or disproved. More recent data, which will be given in the conclusion, show that this problem is apparently more complex and that its solution will require some compromise and coordination not only of these standpoints, but also of some additional considerations.

The first of these explanations of foam stability, put forward by Plateau, is based on the assumption that the persistence of foam films and foams is determined by the slow drainage of the solution. The rate of this hydrodynamic process is obviously determined by the viscosity of the solution in the film, and the role of detergents—foam stabilisers—is thus, on the basis of this theory, to increase the viscosity and consequently to slow down the drainage from the films. As a limiting case, we can imagine a completely "rigid" film with very high persistence.

The validity of these ideas was studied first by Plateau and later by other authors who used more elaborate methods. The so-called surface viscosity of detergent solutions was measured, assuming that the viscosity of the adsorption layer determined the viscosity of the film in which the role of these adsorption layers was particularly prominent. The measurement of surface viscosity, or more correctly, of the surface friction, consists in studying the movement of a body which is semi-immersed in the solution under examination. In a typical experiment, a cylinder is suspended from a torsion filament so that its lower part is immersed in the solution. The cylinder is either made to oscillate around the vertical axis, and the damping decrement of such free oscillations investigated, or the torsion angle resulting from a slow rotation of the vessel containing the liquid (as in Couette's viscometer), is determined. Comparison of these results with those obtained under identical conditions, but in the absence of the detergent, allows a calculation of the contribution of the detergent to the total friction. A correlation between surface friction and foam persistence was indeed found in some but not in all cases. The adherents of Plateau's theory believe that the observed contradictions can be ascribed to the fact that, in view of the non-Newtonian nature of the surface viscosity, this property sometimes does not manifest itself since the velocity of movement is too high in the surface viscosity measurements. Indeed, the velocity of flow in films, especially in very thin films, is very small, and the surface viscosity which is essential for this process should be determined by extrapolating to the zero velocity of measurement; this is very difficult in practice. Furthermore, it is possible that surface viscosity is not related unambiguously to the viscosity in the film, where an additional structure may be present. The most reliable method of ascertaining the occurrence of increased viscosity is stable films would consist in measuring the natural flow in the films. No drainage-induced thinning of microscopic foam films has as yet been investigated in high-stability films. The recent measurements of Lyklema (1962)[18], relating to the spontaneous thinning of large stable vertical films from dodecylsulphate solutions, showed that the rate of solution drainage out of the film corresponds to the normal (not increased) bulk viscosity of the solution. Increased viscosity in the film would of course prolong the period of thinning, and contribute to the extension of foam life, but the fact that it is possible to obtain very stable films without an increase in the film viscosity indicates that this factor is not decisive.

An entirely different concept of the cause of the stability of foam films arose from Gibbs[19] consideration of the "elasticity" of such films. Similar ideas, although in a less clear form, were developed earlier by

References p. 273–274

Marangoni (1871)[20]. It is well worth noting that the relationship between this elasticity and the stability of foam film was suggested by Gibbs with the utmost caution, and with substantial reservations which are often forgotten by his followers.

Gibbs derived the idea of the elasticity of foam films from the following general considerations. One of the characteristic features of thin liquid films is that their thickness is smaller, by many orders of magnitude, than the linear dimensions which determine their area. It may therefore be assumed that in the case of deformations (local expansions and compressions) equilibrium of film components can be achieved in the direction normal to their surface—between the surface and the volume lying below—whereas such an equilibrium has no time to establish itself between individual film sections along the surface, where the distances are considerably greater. Individual film sections may thus be regarded as separate portions, whose volumes and total contents of components do not change during deformations. If, therefore, a certain section of the film (film element according to Gibbs' nomenclature) is stretched so that its surface O increases at constant volume whilst its thickness h is decreased, then since the total content of the surface-active component remains constant, its concentration in the surface will decrease and the surface tension σ will be increased. The increased tension of the stretched section as compared with the adjacent unstretched films will then give rise to a force along the film, which will tend to compress the stretched section and restore uniform thickness throughout the entire film. This force related to a unit increase of the film surface was called by Gibbs the film elasticity:

$$E = 2\frac{\mathrm{d}\sigma}{\mathrm{d}O}O. \tag{8.1}$$

The essential, indeed probably the most essential, feature of these ideas is the fact that the film can counteract such deformations as result in the disturbance of constant thickness along the surface. Any attempts at investigating this phenomenon experimentally or theoretically should therefore be related to the problem of thickness uniformity. In particular, the ideas of Adam[11] referred to before, *i.e.* that in the region of heterogeneous structure of the surface monolayer, where $\mathrm{d}\sigma/\mathrm{d}O = 0$; the Gibbs elasticity should be equal to zero (just as, say, in a one-component film), are wrong. Elasticity should be considered separately for sections covered with a dense monolayer and for sections not covered by such a layer;

this procedure does not necessarily result in the disappearance of elasticity.

The elasticity E introduced in eqn. (8.1) depends on the rate of film deformation and on the total thickness of the layer. The maximum magnitude of E is attained at the fastest deformations, when even the equilibrium between the surface and volume of the stretched section has no time to be established. In such a case the increase of tension is simply determined by a lowering of surface concentration which is not compensated by redistribution from the bulk. In the case of less rapid deformation, involving a wide range of deformation rates, the state of affairs is as described before, *i.e.* the stretching is associated with a lowering of the surface concentration, owing to general redistribution. It is clear in this case that the thinner the film, and therefore the lower the reserve of the surface-active component in the bulk of the liquid, the lesser the extent to which the lowering of surface concentration will be compensated by redistribution from inside, and the higher will be the increase of tension which will tend in the limit to values corresponding to the former case. Finally, if the stretching takes place very slowly, redistribution will extend to sections adjacent to the stretched portion, reducing the differences in their tensions. Such differences, and also the elasticity, disappear at infinitely slow deformations. For the region of medium-rate deformations the function E can be calculated by making use of the conditions that the deformed section preserves constant volume:

$$V = Oh = \text{const.}, \tag{8.2}$$

and a constant content of the surface-active component:

$$cV + 2\Gamma O = \text{const.}, \tag{8.3}$$

where c and Γ are its volume and surface concentrations respectively. It is further assumed for this region that an equilibrium relation is maintained between c and Γ, so that according to the Gibbs adsorption isotherm

$$\Gamma = -\frac{c}{RT}\frac{d\sigma}{dc}. \tag{8.4}$$

To find the final solution it is necessary to know one more relationship between c and Γ or between σ and c. To this end we shall make use o·

References p. 273–274

Langmuir's equation

$$\Gamma = \Gamma_\infty \frac{c}{c+B}, \tag{8.5}$$

[where $B = 1/K$ from eqn. (4.40′)] or, of Szyszkowski's equation, in which case the problem is limited to the surface-active substances which obey this equation (*cf.* Chapter IV). After substituting Γ from (8.5) into (8.3) and differentiating with allowance for (8.2), we obtain

$$\mathrm{d}c = -\frac{2\Gamma_\infty c(c+B)}{2\Gamma_\infty B + h(c+B)^2}\frac{\mathrm{d}O}{O}.$$

From (8.4) and (8.5)

$$\mathrm{d}\sigma = -\frac{\Gamma_\infty RT}{c+B}\mathrm{d}c,$$

which according to (8.1) gives

$$E = \frac{4RT\Gamma_\infty^2 c}{h(c+B)^2 + 2\Gamma_\infty B} \tag{8.6}$$

As expected, eqn. (8.6) shows that E increases with decreasing h and that the relation $E = E(c)$ passes through a maximum at

$$c_{\text{opt}} = B\sqrt{1 + \frac{2\Gamma_\infty}{Bh}}.$$

For the lower homologues of fatty acids and alcohols, to which Szyszkowski's equation can be applied, the calculation discloses that for $h > 10^{-6}$ cm, the second term under the root sign can be disregarded, so that maximum E corresponds to concentrations in the neighbourhood of B. This points to the fact that the Gibbs elasticity might play a stabilising role in unstable foams, where according to Bartsch (Table VIII, 1), maximum stabilities are indeed observed at surface-active component concentrations of the order of B.

The significance of this result should not, however, be overestimated, since in the most interesting region of deformation at medium rates the elasticity of the film should, according to this theory, increase with decreasing thickness.

In actual fact, however, the natural evolution of films is thinning and subsequent bursting when a certain sufficiently low thickness is attained. Since the elasticity increases with progressive thinning, it cannot explain directly the breakdown of films. In view of this fact, Gibbs[21] assumed that the bursting of films is determined by their peculiar properties, which arise in very thin layers. Although this remark is rather vague, it shows that the origin of instability, and consequently of stability, should be sought not in elasticity but in the special properties of thin layers. As has already been shown in Chapter VI, this approach does in fact lead to an explanation of the phenomenon of film destruction.

A further point which may be added is that the ability of a film to maintain a constant thickness can be explained without recourse to elasticity, simply by the fact that any local variation in the film thickness is unavoidably associated with the formation of a deflection in its surface, bringing into play capillary forces which tend to restore the initial state of flat surface and uniform thickness.

It follows from the above that Gibbs' elasticity of the film does not play a decisive part in the mechanism of foam stability, although it is undoubtedly a real effect resulting from our ideas of adsorption.

The third attempt at putting forward a general mechanism for the stability of foams is that propounded by Deryagin and Titievskaya (1953)[15]. In this approach it is assumed that the stability of foam films is due to a positive disjoining pressure which inhibits the thinning process. This suggestion is highly tempting because the effect is easy to visualise, and because it reduces the problem of foam stability to the same factors which effectively explain the stability of lyophobic sols in numerous instances. It is, however, contradicted by a number of considerations. According to the present theory of special thermodynamic properties of free thin layers, the only source of positive disjoining pressure in the case of dilute aqueous solutions of detergents and electrolytes can be Π_{el}. This magnitude is, however, so very considerably decreased in the presence of small electrolyte concentrations that the stability of foams should be sharply reduced by small additions of electrolytes (particularly of multivalent ions), just as in the case of lyophobic sols. Such sensitivity to electrolytes is detected neither in the low-stability nor in the high-stability foams. As regards the possibility of the generation of $+\Pi$ of another more complex type, for instance due to a certain structure of the liquid near its phase surface, direct measurements carried out by the present author with microscopic films of various stabilities have failed to

References p. 273–274

yield any such indications. The only instance (concentrated aqueous solution of butyric acid) when $+\Pi$ of non-electrostatic origin (*i.e.* not from the diffuse electrical layers) was the undoubted cause of enhanced foam stability, represents so far an isolated phenomenon in which the stability of the foam was relatively low (of the order of 1 min). Deryagin's mechanism might perhaps prove more adequate for the non-aqueous foams obtained from concentrated solutions, which are not of particularly high stability.

Let us now pass to a more detailed analysis of the problem of foam stability. We shall at first confine ourselves to the attempt of explaining only one feature of the phenomenon of foam formation, which is most clearly expressed, *i.e.* to the fact that there are two ranges of stability differing considerably in magnitude.

The considerations developed in Chapter VI, which showed that on attaining a certain critical thickness a thin liquid layer becomes unstable with respect to its shape owing to the compensation by the negative disjoining pressure of the effect of capillary forces tending to maintain plane-parallel form and uniform thickness, bring us to the following alternatives: (*a*) the layer bursts on reaching a critical thickness, and (*b*) a sudden change of thickness with the formation of black films takes place when this critical thickness is attained. In the former case foam breakdown takes place, whereas if (*b*) is true, attainment of critical thickness does not destroy the foam, so that the stability is considerably higher, depending on the strength of the black films. The transition to high stability, for instance when the concentration of the detergent-stabiliser is increased, should therefore be accompanied by the formation of black spots, which should not form in films of low stability.

These ideas are so well borne out by experimental evidence[6] that it is reasonable to analyse the practical results in greater detail. It has already been mentioned that a transition from low to high stability can be observed with many detergents [S-shaped curves of $\tau(c)$ in Fig. 63]. Simultaneous observation of the formation of black spots was carried out with the same solutions on microscopic foam films in reflected light. To eliminate the effect of Π_{el}, which stabilises the microscopic films without having a noticeable effect on the stability of foams, the solutions used in all measurements [in studies of foam persistence $\tau = \tau(c)$, in observations of microscopic films, and in determinations of the isotherms $\sigma = \sigma(c)$] always contained the same amount of electrolyte, *i.e.* 0.1 moles KCl per litre*. Observations carried out with microscopic films revealed the following effects: At low concentrations of the detergent, corresponding to low foam stability of the order of 10 sec, the microscopic films thin out relatively rapidly after their formation and burst suddenly at a certain thickness h_{cr}, of the order of 200 to 300 Å, without forming any black spots. The behaviour of these films is exactly the same as that of films formed from the solutions of low-molecular foaming agents such as the lowest fatty acids and alcohols. In the concentration region, where the steep rise of the $\tau = \tau(c)$ curve is first observed, black spots form in the microscopic films prior to bursting. The minimum concentration c_{bl} at which these spots appear in the shape of black dots immediately before the bursting of the film is highly reproducible. When the detergent concentration is increased further, the black dots have sufficient time to grow to larger dimensions before bursting, and finally merge to a continuous microscopic black film which has in many cases an infinite stability.

The above described correlation between the steep increase of foam persistence with concentration of the detergent on the one hand, and the onset of the formation of black spots on the other, was obtained by Ekserova (1962–1963) for all investigated detergents, *i.e.* for:

OP*n* (octylphenol polyglycol ethers), where $n = 7, 9, 15$ and 20,
alkyl sulphates RSO_4Na ($R = C_nH_{2n+1}$, where $n = 6, 8, 10, 11, 12$, and 14,
octyl sulphonate $C_8H_{17}SO_3Na$,
commercial foaming agent Igepon T (methyl tauride of oleic acid).

It is still uncertain why black spots appear at the concentration c_{bl} and why they are absent in the presence of low-molecular foaming agents. The present author's suggestion that the origin of black spots is related to the properties of the adsorption layer was checked by parallel measurements of the dependence of surface tension σ on concentration, finding that at the concentration c_{bl}, a discontinuity is always observed in the $\Delta(c)$ curve, with a small horizontal section which is clearly defined in the

* The electrolyte concentration was so chosen as to make the equilibrium thickness of the microscopic foam film less than the critical thickness of rupture or of the formation of black spots — in general of the thickness at which the film loses its shape stability.

References p. 273–274

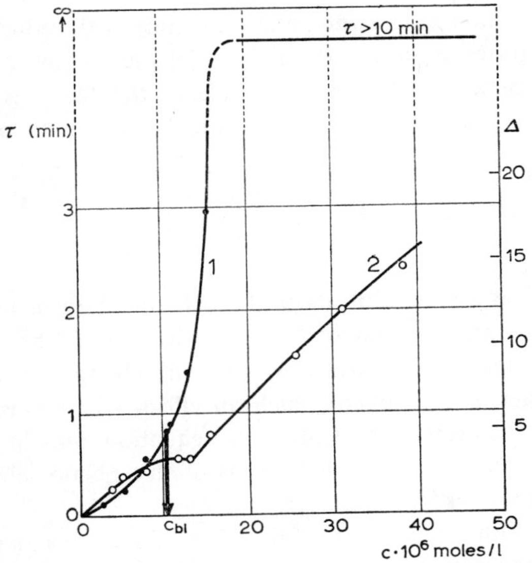

Fig. 65. Aqueous solution of Igepon T of concentration c (molecular weight about 390) + 0.1 mole KCl per litre. (1) Dependence of foam persistence τ on concentration. (2) Dependence of the lowering of surface tension Δ on concentration. c_{bl}–concentration at which black spots begin to form.

case of detergents of higher molecular weight. The occurrence of this horizontal section points to the fact that black spots arise in the adsorption region where the dense adsorption monolayer begins to form. Such a curve was shown in Chapter IV (Fig. 26). Fig. 65 shows the curves of $\tau(c)$ and $\Delta(c)$ (where Δ is the lowering of surface tension in relation to the tension of the solution at $c = 0$) and the concentration c_{bl} for Igepon T; these curves are a good illustration of the considerations described above. It remains unclear, however, just which property of the dense adsorption monolayer of the detergent determines the possibility of the formation of black spots.

The above results allow the following summary of the problem of foam stability:

1. *Low-stability foams with a life ranging from several seconds to several tens of seconds form from dilute aquoeus solutions of detergents* ($c < c_{bl}$) *and from solutions of low-molecular surface-active substances.* The persistence of such foams is determined by the time of their thinning down to the critical thickness h_{cr}. This time is particularly sensitive to h_{cr}, since

the process of thinning slows down considerably with reduction in film thickness. The proposition that the stability of such foams is determined by the Gibbs elasticity is not very probable, as this factor is not related to the drainage of solution from the film. According to our formula (6.21)

$$h_{cr} = \sqrt[4]{\frac{3k\lambda^2}{64\sigma}},$$

if we suppose that the concentration of the foaming agent in this range of stability exerts an effect on λ, then this effect should be shown by a change in the length of the surface wave λ with changing concentration of the surface-active component. Such an effect, which was not taken into account in the derivation of the above equation, may in fact occur; the foundation for a quantitative analysis of this phenomenon may be found in Levich's book[22].

In agreement with general ideas of the effect of surface-active agents on surface waves, adsorption increases wave damping and at the same time reduces the length of the surface waves; this effect should pass through a maximum in the region of concentrations approximating to the constant B in Szyszkowski's equation. If such an effect really exists, and this could be established experimentally, then h_{cr} should also pass through a minimum at $c \approx B$, so that τ would exhibit a maximum in this range of concentration. This explanation of the maxima in $\tau(c)$ curves obtained for solutions of low-molecular foaming agents is so far merely hypothetical. To calculate $\tau(c)$ it is necessary to know the hydrodynamic law of the thinning of foam films, which is at present known only for very thin films with a very small diameter. The problem is still unsolved and is apparently highly complicated for the relatively large films which give rise to foams. Future investigators of this complex problem may perhaps be consoled to a certain extent by the fact that in the region of low stabilities the average critical thicknesses of rupture are so large* that in this case it is permissible to disregard in the draining problem the special properties of thin layers (the disjoining pressure) and possible rheological effects. This is because the films are fairly thick, and because in this case

* The average critical thicknesses determined conductometrically by the present author (1957)[23] for large films with an area of the order of 1 cm^2 were found to be of the order of 1 μ, *i.e.* roughly 100 times greater than the thicknesses of the microscopic films formed from solutions of the same composition[6].

References p. 273–274

we can confine ourselves to low-molecular foaming agents which do not give rise to rheological effects.

2. *High-stability foams with a life ranging from tens of minutes to tens of hours, which form from solutions of detergents when the concentration is $c > c_{bl}$.* In this case the period of thinning is of course short compared with the persistence of films of a finite constant thickness, so that foam persistence is now determined by the time of the existence of such films. Most studies are concerned with this type of foam, which is partly due to the fact that foam properties of practical importance, *e.g.* their detergency, are optimum at foaming-agent concentrations far in excess of c_{bl}, close to the concentration at which micelles are formed. At such high concentrations of the foaming agent the special characteristics of the mechanism leading to the production of high-stability are, however, completely concealed. This is probably the reason why the phenomenon of black spots and the mechanism of their formation (corresponding to c_{bl}) have been ignored until recent times. The mechanism of the spontaneous breakdown of constant-thickness foam films, which is probably of a fluctuation type, has not as yet been studied except for the special case of bimolecular Perrin films whose mechanical strengths were analysed quite recently by Deryagin and Gutop[17], on the basis of the cavitation-hole mechanism of rupture. Unfortunately, even in this case, concrete calculations require the knowledge of certain constants which may only be determined by further investigations. The design of experiments involving the use of artificial, known and controllable mechanical stimuli which disrupt such films will probably be decisive in the elucidation of this complex problem of considerable practical importance. It will also enable the detergents to be classified according to their effectiveness as foam stabilisers.

3. Greatest complexity is encountered in the study of foams in the transition region of stability, when the detergent concentration is $c \approx c_{bl}$. Such studies are highly important. In this case, the period of thinning and the life of the constant-thickness thin films are of the same order, the persistence increasing gradually with increasing concentration of the detergent. This ensures a gradual, although admittedly steep, transition from low to high stability. It is clear that in this case any rheological effects in the films may play an important part, by prolonging the period of thinning. Moreover, since in this region (at $c = c_{bl}$) there arises a new, more powerful mechanism of foam stability, it is a convenient region in which to study its nature, *i.e.* the nature and the part played by black spots.

In conclusion, let us consider the method of investigating large flat foam films obtained by withdrawing a suitable frame from a solution of detergent. Owing to the application of optical thickness measurements and of a number of other ingenious devices, this old method has recently been revived (in the experiments carried out by Mysels[24], Overbeek[25], Duyvis[14] and Lyklema[18]) and is showing further promise. The main disadvantage of the frame method, common in fact to all studies of large films, is that only very stable films can be used. It has already been mentioned that this method makes it possible to establish the presence or the absence of rheological phenomena in a film. Moreover, it allows an observation of the mutual displacement of thin and thick areas in the film, and of the highly complex processes which take place near the thickened film ends, which apparently play an important role in the general behaviour of such films. In this way, Mysels succeeded in demonstrating in an elegant way the phenomenon of "stiffening" at a definite composition and surface concentration of the stabiliser. By lowering and raising another frame in the solution, it is possible to vary rapidly the total surface and thus control the surface concentration of the film in the stationary frame. Furthermore, by changing the surface concentration a transition from a "mobile" to a "rigid" film may be accomplished. In general, large films in principle allow direct measurements or evaluations of the surface concentration to be effected, which is obviously of great importance. Thus for instance, by blowing a soap bubble from a definite volume of solution of a given concentration, Sheludko and Polikarova (1954)[26] succeeded in determining (from the critical bubble volume at which it burst) the critical breakdown thickness and the critical surface concentration of soap, at which the film was ruptured. It was found that the critical concentration is similar to (slightly less than) the saturation concentration of the monolayer. This admittedly isolated result is in agreement with the ideas developed earlier concerning the special stabilising effect of dense adsorption layers of the detergent.

4. Emulsions

Heterogeneous disperse systems known as emulsions occur widely in nature (*e.g.* milk) and in industry (numerous lubricants, a number of foodstuffs such as margarine, cosmetic creams, etc.). Apart from the conventional meaning of this word (dispersion of a liquid in a liquid),

References p. 273–274

the light-sensitive suspensions of silver-halide in gelatin are incorrectly termed emulsions in the field of photography.

Artificial emulsions are usually obtained by the dispersion method, *i.e.* by stirring sufficiently vigorously a mixture of two immiscible liquids so that drops of one phase are formed in the other phase. Being liquid, these drops are elongated in the direction of stirring, up to a certain degree beyond which the drops lose their stability and undergo comminution*, increasing the degree of dispersion. This comminution is opposed by the increasing probability of coalescence with increasing number of droplets, so that a steady-state condition is finally established, which is characterised by a certain maximum degree of comminution and by a certain drop size distribution. This limiting state depends largely on the presence of stabilisers which prevent coalescence, the so-called emulsifiers. In dilute emulsions an increase in the degree of dispersion leads to increased stability, owing to the reduction of the rate of sedimentation or creaming. Thus, when milk is to be transported for some time, the separation of cream is prevented by additional dispersion ("homogenisation"). Various types of mixers, *e.g.* the Plauson colloid mill, are used to obtain emulsions by mechanical methods.

In stable concentrated emulsions (creams) the concentration of the disperse liquid can attain very high values. It may be shown from purely geometrical considerations that, if the drops are monodisperse, the maximum condensation of spherical particles without deformation corresponds to a volume concentration of about 74%, irrespective of particle size. Considerably more concentrated emulsions may be obtained in practice, provided they are polydisperse, and with the aid of drop deformation and the formation of a foamlike structure. Volume concentrations of the disperse substance can approach 100% in such emulsions.

A highly interesting characteristic of emulsions consists in that, when two mutually insoluble liquids are mixed, phase 1 can be dispersed in phase 2 or phase 2 in phase 1. A special convention for denoting the type of emulsion has therefore been adopted. One of the phases constituting the emulsion is usually water or an aqueous solution, and this phase is denoted by "W". The other non-aqueous liquid is denoted by "O" and is called "oil", irrespective of its actual nature. The emulsion of water in

* According to the theory of capillarity, a liquid cylinder becomes unstable when its length becomes greater than the circumference of the cross section perpendicular to the axis.

benzene is *e.g.* denoted by W/O, and that of benzene in water as O/W.

It is of interest to analyse the causes underlying the formation of a given type of emulsion.

In their foreword to the Russian edition of Clayton's book[1], Rebinder and Pospelova summarise the material available in this field as follows:

During the mechanical mixing of two mutually insoluble liquids both W/O and O/W emulsions are formed simultaneously, and the emulsion which corresponds to the dispersion of the liquid with a lesser volume has a smaller concentration. Other conditions being equal, this more dilute emulsion will therefore be more stable, owing to the lower probability of collisions between its particles. The probability of a collision between particles is roughly proportional to the concentration, and the concentration ratio of two emulsions is roughly proportional to the volume ratio of the emulsified liquids. The stability ratio of the two emulsions is thus roughly inversely proportional to the volume ratio of the liquids used for the preparation of the emulsion. Furthermore, the final result of the emulsification can be predicted since the more stable of the two emulsions will "survive".

Various stabilisers exert unequal effects on the effectiveness of the collisions in W/O and O/W emulsions. It is also necessary, therefore, to consider the effectiveness of collisions leading to coalescence. The effectiveness of collisions is measured by the time τ for which two drops of the same liquid, separated by a layer of the other liquid, can exist under identical conditions without merging with each other.

If the ratio of volume is multiplied by the ratio of lifetimes defined as above for the two possible emulsions, a measure of the relative stabilities is obtained, which takes into account the probability of collisions and the probability of merging, *i.e.* both factors which determine the rate of coalescence. If this criterion is considerably greater than 1, or in other words if the direct emulsion is much more stable than the reverse one, it may then be stated that the final product will favour the more stable emulsion.

The situation becomes more complex when the above ratio is close to unity, *i.e.* when the two possible emulsions possess similar stabilities. In this case, an appreciable influence is exerted by the walls of the vessel and by other objects (*e.g.* the stirrer) with which the mixture comes into contact. If these are for example wetted by one liquid but not by the other, then destruction of the emulsion of the wetting liquid is promoted on the solid surfaces, and the wetting liquid tends to become the dispersion medium. This factor can predetermine the type of the final emulsion.

The so-called inversion of an emulsion is closely related to the problem in question. This phenomenon may be observed when a substance is added to the system, which alters the properties of the stabiliser in such a way that it exerts a more stabilising effect on the inverse emulsion.

The type of emulsion may be determined by several methods.

1. The microscopic method. In this procedure, a dye which preferentially colours one of the phases is added to the emulsion. If the preferred phase is known in advance, microscopic examination will readily determine which phase has been dispersed. The disadvantage of this method consists in that the new component introduced into the system, may frequently be surface-active and affect the type of the emulsion.

2. In another very simple and convenient method the emulsion is successively diluted with both phase components. An emulsion will readily mix with the phase which constitutes its dispersion medium, but much less easily with the other phase which is dispersed in the emulsion. This procedure is characterised by the same disadvantage as the microscopic method, although no third component is introduced in this case. This is because, as has already been mentioned, the type of emulsion which forms on the mixing of two phases depends on their volume ratio. By adding one of the phases we create a considerable local increase of that component, which in some cases, causes the emulsion to invert.

3. The above disadvantages are avoided in the conductometric method, which is particularly reliable when the two phases differ considerably in their conductances (which is frequently the case). If the emulsion is of the O/W type, *i.e.* if the organic liquid is dispersed in the aqueous solution, the measured conductance of the emulsion is due to the conductance of the aqueous solution which ensures a continuous contact between the electrodes of the conductivity cell. In the case of a W/O emulsion, the conducting water droplets are surrounded and isolated by a non-conducting organic liquid so that the measured conductance is close to zero.

The conductometric method fails to give accurate results only in the case of highly concentrated stable emulsions in which complex systems of the two emulsion types are formed, or if strong electric fields are used, which can break through the very thin films of the non-conducting phase, separating the conducting droplets of the dispersed phase*. The conducto-

* This is utilised in breaking down the emulsion of water in petroleum.

metric method is very convenient for following the inversion process of an emulsion.

Considerably more complex structures of emulsions are known to exist. The disperse drops may themselves be emulsions by occluding smaller droplets of the dispersion medium. This "internal" complication of the structure may be not confined to one degree, *i.e.* the occluded droplets may also be emulsions.

The presence of a third component (emulsifier) is necessary for the formation of concentrated emulsions, just as the presence of foam stabilisers is required for the formation of foams. Moreover, even as far as their nature is concerned emulsifiers closely resemble the foaming agents. Both are substances of medium molecular weight characterised by strong adsorption at the interface in question. Soaps possess an exceptional stabilising effect on both foams and emulsions.

It has already been mentioned that the theory of emulsions and of their stability is even less developed than the theory of foams. According to the earliest point of view, a sufficiently low surface tension at the interface between the emulsion-forming phases is a necessary condition of stability, just as in the case of foams. In such a case the excess surface energy of the disperse system is low, and the thermodynamic tendency directed towards lowering the energy of the system (reduction of total surface area) and the degree of dispersion, is also reduced. These ideas, which reduce the stability to elementary thermodynamic factors, are entirely incorrect when applied to foams, and that is why they have not been analysed in that sector. In the case of emulsions they are somewhat better substantiated, since the tension at the interface between two liquids is either close to that at the critical mixing point, or can be made very small by the addition of surface-active agents. The dispersions which form spontaneously under such conditions, and the causes of their formation have already been discussed in Chapter IV.

The non-spontaneous formation of emulsions in the presence of emulsifiers is considerably more important and is more frequently observed. The non-spontaneous emulsions resemble foams, and the origin of their stability must be ascribed to more complex factors. Ideas, similar to Plateau's theory of foam stability, concerning the role of the mechanical strength of thin layers separating the drops in a concentrated emulsion, are fairly widely accepted. The concept of mechanical strength, widely used in the research of Rebinder and his school, is not very clear, and it is

References p. 273–274

therefore difficult to understand the essence of these concepts. In the simplest case, when we consider the increase in film viscosity brought about by the emulsifier, then as in the case of foams the problem is reduced to the retarding mechanism of the thinning of emulsion films, which can be ascribed in emulsions to the same factors as in foams. Confirmation of this mechanism is fairly complicated. In view of the absence of systematic model studies of the thinning of emulsion films, it is at present difficult to make with confidence any definite statements on this subject. Nevertheless, by analogy with foams and from the available data relating to foams, it is possible to assume that this mechanism does not play a decisive role. On the other hand, if the mechanical strength of a film is taken as the totality of "mechanical" properties, including the still uncertain mechanical properties of the adsorption monolayer, which hinder the rupture of the thin layer, then—again proceeding from the analogy with foams and with the a priori considerations expressed in the discussion of foams—we can assume that the rate of coalescence in emulsions is controlled by similar factors. Unfortunately, in view of the absence of results relating to the mechanism of the thinning and rupture of emulsion films, it is at present impossible to go beyond such fairly indefinite assumptions.

The similarity between emulsifiers and foam stabilisers is not accidental. As a matter of fact, both in emulsions and in foams (considering concentrated dispersions), the problem of the stability of a disperse system is reduced to the problem of the stability of thin liquid layers which separate the dispersed phase. The emulsions are, however, characterised by the fact that in this case the thin layers separate a condensed phase, whilst in foams they border on gas. This complicates a theoretical discussion of the problem to a certain extent. Since even in the relatively simpler case of foams the causes of stability have not as yet been fully revealed, it is understandable why this problem remains particularly unclear in the case of emulsions.

In the absence of any theory, the emulsifiers are chosen by purely empirical methods, and since the practical importance of emulsions is tremendous, such experimental investigations and the various recipes discovered are particularly numerous and are the subject of many patents. Understandably, the numerous studies of emulsions have given rise to a large number of explanations for the mechanism of the action of emulsifiers, these explanations being mostly qualitative, often contradictory, and incapable of wider generalisation over the available experimental material. All of these explanations cannot be quoted, in view of their multiplicity, nor would there be much point in such a list in view of their incompleteness. The state of the problem will, however, be illustrated with a few examples.

One of the characteristics of emulsifiers, *i.e.* their surface activity, is generally known. For an emulsifier to be surface-active at the interface, *i.e.* to be adsorbed on it, it should possess comparable affinities to both phases. If this condition is not fulfilled, the emulsifier will be "drawn in" from the interface into the phase to which it exhibits greater affinity. Typical examples of such "balanced" surface-active substances are found among some soaps, the molecules of which have an organic "tail" with a strong affinity towards the "O" phase, and an ionized "head" with a considerable affinity to the "W" phase. When the "tails" are very short (lower soaps) the affinity to "W" predominates, and the molecule is drawn into "W". Such an emulsifier is regarded as "unbalanced". When the "tails" are very long, the affinity to "O" is predominant, and the emulsifier is again poorly "balanced" but as it were in the other direction. A definite mean relation between "head" and "tail" should therefore result in the optimum "balance", and consequently in the optimum stabilising effect.

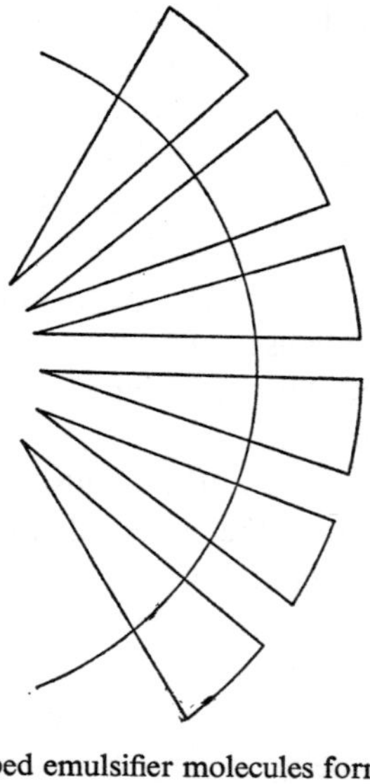

Fig. 66. Wedge-shaped emulsifier molecules form a curved interface.

References p. 273–274

This is indeed the case: the best emulsifying properties are found in the salts of fatty acids of medium molecular weight, such as oleates and stearates.

The explanation of the emulsifier property of producing one type of emulsion is of great interest. It was found that after slight changes in the composition, certain emulsifiers produced different types of emulsion and led to inversion. Thus the alkali metal salts of medium molecular weight fatty acids yield O/W emulsions, whilst the same salts of divalent metals (*e.g.* Mg) yield W/O systems. If the concentration of divalent ions in an O/W emulsion stabilised by a soap with a univalent metal cation is gradually increased, the emulsion undergoes an inversion and passes into the W/O type; both types are unstable in a certain range of concentration.

In 1917 Harkins, Davis and Clark[27] tried unsuccessfully to explain the relation between the properties of emulsifiers and the types of emulsion they stabilised. These authors tried to explain this phenomenon by the "wedge-like" shape of the emulsifier molecules. If the emulsifier molecule is in fact wedge-like, it is clear that the interface should be curved (Fig. 66), and the disperse phase will consist of the component with respect to which the interface is concave. This theory cannot, however, stand up to a quantitative check (Adam[11]). When allowing for the geometrical shape of the stabiliser molecules, it was expected that for a given pair of phases, when the emulsions were stabilised by fatty acids with a chain length equal to ~ 100 Å, emulsions with drop size of the same order of magnitude would remain. It was found in fact that droplets of stable emulsions were greater by a factor of many hundreds.

A qualitative but more general theory of the action of emulsifiers was suggested in 1940 by Schulman and Cockbain[28]. According to this theory, special significance is attributed to the second component of the emulsifier, which is more soluble in one of the phases. Owing to the effect of this second component, a definite structure of the thin layer is formed, which is of decisive value for its stability.

Any explanation of the observed properties of emulsifiers should apparently be based on the special properties of thin liquid layers. It can therefore be expected that this problem will be solved when the special properties of these liquid layers are finally elucidated.

In conclusion, let us consider the model investigations carried out by Sonntag (1962)[29], which demonstrate the remarkable similarity of foams and emulsions. Two tubes filled with mercury are placed vertically above each other in a non-polar liquid (xylene, octane), to form a free mercury

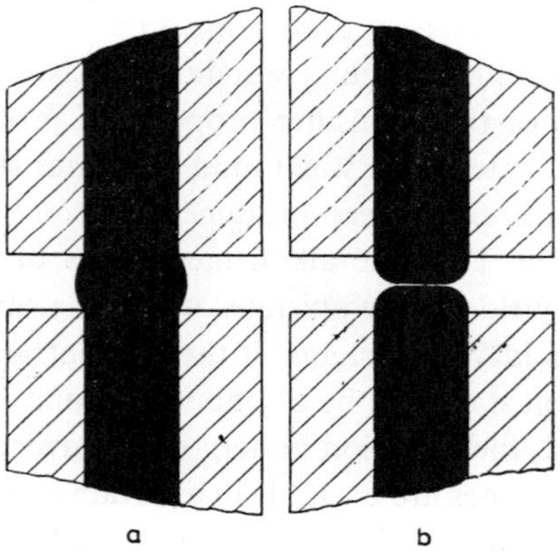

Fig. 67. (a) Free vertical mercury column. (b) Thin layer compressed between mercury menisci, after the column has been cut through.

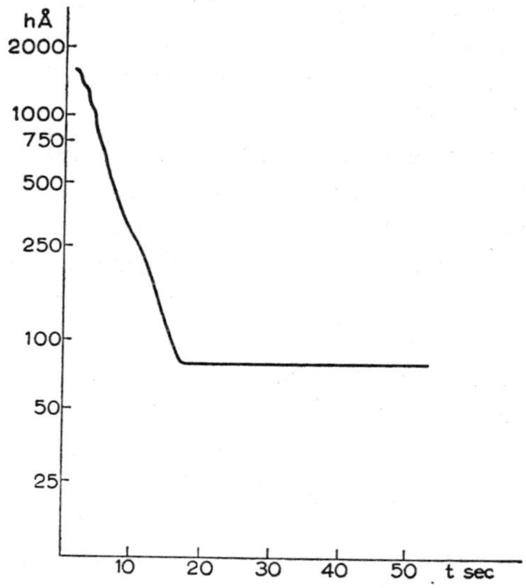

Fig. 68. Dependence of thickness h on time for a xylene film fixed between mercury menisci. Stabiliser—10^{-3} mole/l of stearic acid.

References p. 273–274

column in the intervening gap (Fig. 67a). This column is cut through with a mica knife. If the system contains an emulsifier (*e.g.* stearic acid), a thin layer of the non-polar liquid is formed, compressed between the two mercury menisci (Fig. 67b*). By measuring the capacitance between the two parts of the broken mercury column, and by determining the film diameter by lateral observation, data are obtained from which the thickness of the thin layer and its time dependence may be calculated. This method of measurement makes it possible to record in a very simple and convenient manner the development of the film thickness with time. The results obtained by this procedure are in excellent agreement with the data relating to free foam films; in other words, when the concentration of the emulsifier is low and consequently the film persistence (evaluated statistically as "half-life" $\tau_{\frac{1}{2}}$) is short, one observes a smooth thinning out of the film, which ends in rupture. When the concentration of the emulsifier (stearic acid in xylene) is higher, the thinning process stops at a thickness estimated as about 70–80 Å (Fig. 68), and a film of constant thickness is formed, whilst the life of the film is greatly extended. At very high concentrations of the emulsifier rupture occurs again, but this takes place at a much greater thickness, of the order of 1000 Å, and film life is again considerably reduced. The non-thinning film 70 to 80 Å thick is most probably a Perrin film—a double layer of stearic acid. This assumption is also supported by the fact that this thickness is formed at the discontinuity in the thickness–time curve, and not gradually as is usually observed in the case of multimolecular equilibrium films. When studying film thinning by the method used by the present author for free films, Sonntag calculated on the basis of Reynold's law the disjoining pressure in the film. It was found that in this case, too, as in foam films, the disjoining pressure in the multimolecular film is negative, and is of the same order (Fig. 69). It is therefore a de-stabilising factor. The results of a parallel measurement of the surface viscosity of the monolayer are highly interesting. An excellent correlation with stability was found for a xylene layer stabilised with stearic acid (Fig. 70). On the other hand, it is obvious that since direct measurements disclosed the formation of a non-thinning film (implying that the high stability is not associated with the draining of solution), viscosity cannot be the direct cause of stability. Insofar as the surface viscosity is determined

* In the absence of an emulsifier the two columns of mercury merge immediately into one another.

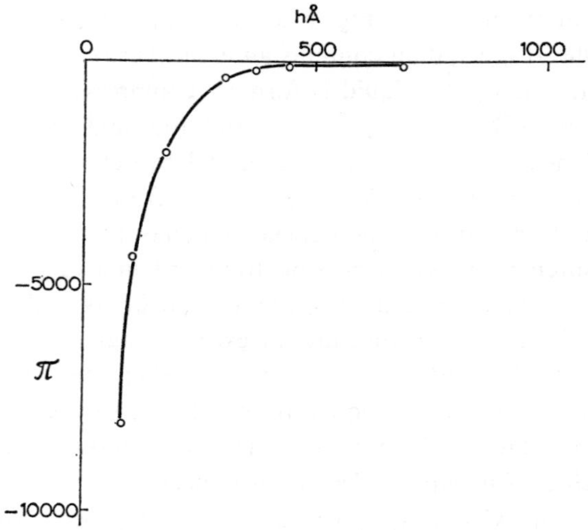

Fig. 69. Isotherm of the disjoining pressure $\Pi(h)$ of a xylene layer ($+1.8\cdot10^{-3}$ moles/l of butyric acid) fixed between mercury menisci.

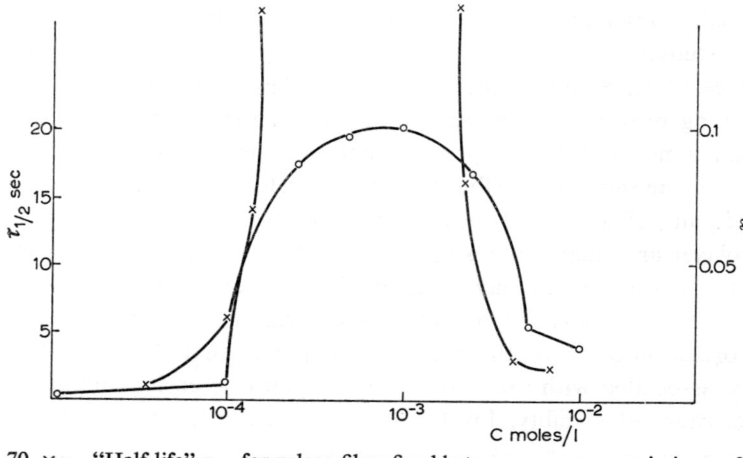

Fig. 70. × — "Half-life" $\tau_{1/2}$ for xylene films fixed between mercury menisci, as a function of stearic acid concentration. ○ — Surface viscosity $\eta_s(c)$.

References p. 273–274

by intermolecular adsorbate/adsorbate adhesion forces which are also associated with other mechanical properties of the monolayer, the occurrence of a correlation between surface viscosity and the stability is possible, but not indispensable. Indeed, just as with foams, Sonntag failed to detect any correlation between the surface viscosity of an emulsion and its stability in other cases.

The above results represent the first step in model investigations of emulsions. The convenient method of thickness measurement used by Sonntag is still insufficiently reliable, mainly because of the probable great difference in the dielectric constants of xylene and of the adsorption monolayers of the film, but also because it is impossible to control the profile of the film. The isotherm $\Pi(h)$ given in Fig. 69 cannot therefore be regarded as final. The capacitance method of thickness determination resulted in the selection of a highly unusual mercury–oil system, with which it is difficult to investigate the stability of the emulsion, since an emulsion of mercury in oil is extremely unstable owing to the pronounced sedimentation caused by the high density of mercury. In spite of the above drawbacks, the results obtained illustrate so clearly the possibilities of such methods and the similarity between emulsions and foams, that we can look forward with confidence to a further development of similar model studies of emulsion systems.

REFERENCES

1 CLAYTON, *The Theory of Emulsions and their Technical Treatment*, 4th ed., London, 1943.
2 BIKERMAN, *Foams; Theory and Industrial Application*, Reinhold, New York, 1953.
3 DE VRIES, *Foam Stability*, Rubberstichting, Delft, 1957, Comm. 326.
4 KITCHENER AND COOPER, *Quart. Revs.*, 13 (1959) 71.
5 MYSELS, SHINODA AND FRANKEL, *Soap Films; Studies of their Thinning*, Pergamon, New York, 1959.
6 SHELUDKO, *Proc. Koninkl. Ned. Akad. Wetenschap.*, B 65 (1962) 76.
7 BARTSCH, *Kolloidchem. Beihefte*, 20 (1924) 27.
8 HARDY, *Proc. Roy. Soc.* (*London*), A 86 (1927) 27.
9 TALMUD AND SUKHOVOL'SKAYA, *Z. Phys. Chem.*, 154 (1931) 277.
10 TRAPEZNIKOV, *J. Phys. Chem.* (*USSR*), 14 (1940) 821.
11 ADAM, *The Physics and Chemistry of Surfaces*, 3rd ed., Oxford University Press London, 1941.
12 PERRIN, *Ann. Phys.* (*Paris*), 10 (1918) 160.
13 WELLS, *Ann. Phys.* (*Paris*), 16 (1921) 69.
14 DUYVIS, Thesis, Utrecht, 1962.

15 Deryagin and Titievskaya, *Kolloidn. Zh.*, 15 (1953) 416; *Proc. 2nd Intern. Congress of Surface Activity*, (1957) 211.
16 Dewar, *Proc. Roy. Inst.*, *Vol. 22*, 179, 359; *Vol. 24*, 197 (1917–23).
17 Deryagin and Gutop, *Kolloidn. Zh.*, 24 (1962) 431.
18 Lyklema, *Rec. Travaux Chim., Pays-Bas*, 81 (1962) 890.
19 Gibbs, *Collected Works*, Longmans Green and Co., New York, 1928, Vol. 1, *Thermodynamics*, p. 300.
20 Marangoni, *Il Nuovo Cimento*, 516 (1871/72) 239.
21 Gibbs, *Collected Works*, Longmans Green and Co., New York, 1928, Vol. 1, *Thermodynamics*, p. 305.
22 Levich, *Physico-chemical hydrodynamics*, Prentice-Hall, Englewood Cliffs, 1962.
23 Sheludko, *Z. Elektrochem.*, 61 (1957) 220.
24 Mysels, Cox and Skewis, *J. Phys. Chem.*, 65 (1961) 1107; Mysels and Cox, *J. Colloid Sci.*, 17 (1962) 136.
25 Overbeek, *J. Phys. Chem.*, 64 (1960) 1178.
26 Sheludko and Polikarova, *Annuaire Univ. Sofia*, 49 (1954/55) 15.
27 Harkins, Davies and Clark, *J. Am. Chem. Soc.*, (1917) 595.
28 Schulman and Cockbain, *Trans. Faraday Soc.*, 36 (1940) 651, 661.
29 Sonntag, *Z. Phys. Chem. (Leipzig)*, 221 (1962) 365, 373.

Index